REGINALD MELMOTH GORDON GATACRE

'Gat'

Philip Gatacre

Claverley Publishing

First published in 2020
by Claverley Publishing
185 River Street
Deniliquin, 2710

A catalogue record for this
book is available from the
National Library of Australia

NATIONAL
LIBRARY
OF AUSTRALIA

ISBN 978 0 64682258 7

Design by Skeleton Gamblers Creative

Contents

Reginald Melmoth Gordon Gatacre

PROLOGUE

My father, Reginald Melmoth Gordon Gatacre, was always known throughout his working life as Browser and Gat to his grandchildren, whom he loved. (On his first day as a beginner jackeroo the rest of the group decided he needed a nickname. 'Browser' was their choice, possibly because a browser is a young sprout, or new growth.) He had an exciting and idyllic childhood at 'Boisdale' in Wooroolin in Queensland's South Burnett district.

My grandfather Reginald Henry Winchcombe, and his wife Christian Esson Gatacre had three children, of whom Melmoth, as the family all called him, was the eldest; Galfrey George Ormond, who was two years his junior, and Louisa Sophia Elizabeth, four years younger again. Melmoth was close to his brother Galfrey; they were always playing and doing all things young boys do. After attending primary school in Wooroolin, in 1915 he went to Church of England Grammar School in Brisbane for about four years. As a seventeen-year-old he went back to near Wooroolin to work on a station, 'Cooindah', owned by Dick Tancred, a friend of his father. Then he went on to 'Bombyanna a property at Lalaguli owned by the Parker family who were also family friends. They gave Melmoth his first reference in which they stated that 'this lad has ability, strength and character and I would recommend him'. It was from these two experiences he fell in love with working on the land with horses and all the animals that go with it. He was, however, very sceptical about new-fangled gadgets.

Reginald Henry Winchombe and his wife, Christian Esson Gatacre, Gat's father and mother

After his wonderful time in Queensland he decided to further his experience on a well-known large sheep grazing property, 'Boonoke', owned by the famous Falkiner family about whom he had heard through his cousin. He intended to stay only two years but it was twenty-six years later when he eventually left.

Gat at Cooindah

In 1940 Melmoth married Winifred (Winkie) Stokes who had been the governess for the Falkiners at Boonoke. They were to have three children: Philip Melmoth, William Michael and Winifred Louise. Melmoth was a good-looking and distinguished gentleman and when they became engaged Winkie took him down to Melbourne to meet her sisters. They were very impressed with him and one of them said, 'My, Winkie, you really have a Spencer Tracey.' They knew all the Falkiners very well and they were among his closest friends.

After leaving the Falkiners to make way for their nephew, Travis Falkiner, Dad inquired about purchasing some land, so Otway Falkiner sold him some acres south of Zara; he named this property 'Claverley'. While he had Claverley he still managed Wanganella, a 28,000-acre property, owned by the Austin family. After only a few years illness struck. He lost feeling in his legs and had to go to Melbourne for back surgery to repair a blockage in his spinal cord. This would've been ground-breaking surgery at the time, and although it was successful he was limited in mobility, so he moved his family into town to live.

This would be the start of forty years of semi-retirement but he always wore a woollen tie with either a Viyella or Airtex shirt. He took a keen interest in the merino sheep on Claverley and any other farming that was going on. Also he would do some sheep classing and ram selecting for friends. He enjoyed attending Merino Field Days and local sheep and cattle sales.

Browser served on the local Pastures Protection Board between 1961 and 1975 and was also chairman for a few years.

During the 1980s he had lost partial sight in one eye. Then during the hay fever season he had a bout of coughing which caused an aneurism in the other eye. This left him with limited sight. It was then that he decided to write down his life experiences. This was done in exercise books, on pieces of cardboard and any scraps of paper he could find. They were mostly brief notes which we went through together, using a tape recorder, and he expanded on the notes; my daughter Melinda put them on the computer.

Browser loved his wife Winkie and all his children and his grandchildren; he loved telling them stories from his early years.

However, tragedy would strike Winkie and Browser when, in 1992, their daughter Louise Officer lost her long battle with cancer; they were both devastated.

He lived a long and wonderful life to the age of ninety; he was quietly self-confident, sometimes a little stubborn but always a perfect gentleman. He lived life with a minimum of fuss.

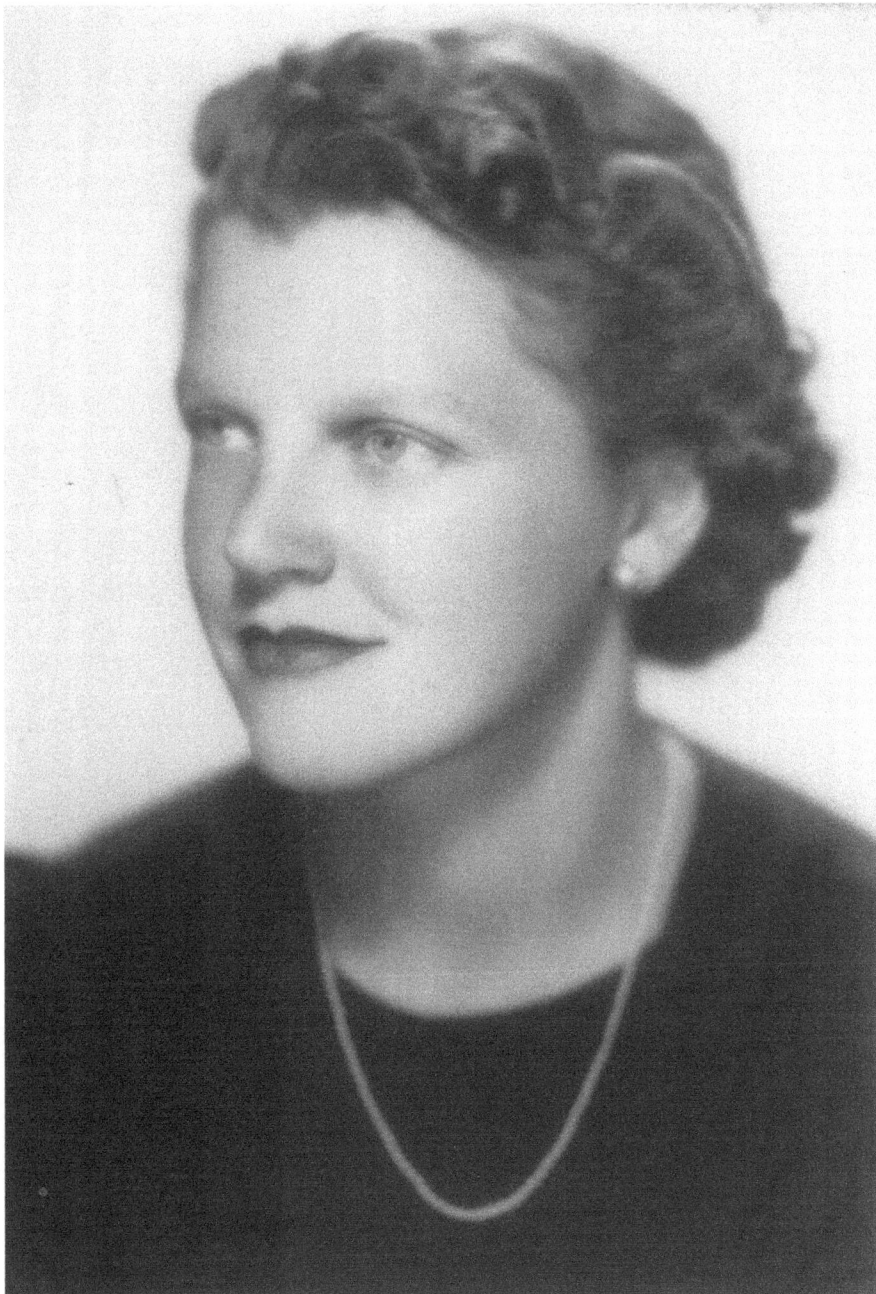

Winifred (Winkie) Stokes,
Gat's wife

Capt Combe   Miss Campbell   Mrs Combe

1 FAMILY HISTORY

Gat's father and grandfather in this picture at the Sydney Lawn Tennis Club, Bath, UK

MY ANCESTORS WERE ENGLISH. Of the families descended from the mediaeval gentry, nine have the proud distinction of being descended in the male line from an ancestor who took his surname from the lands which they still hold. One of the nine is the Gatacre family.

The name Gatacre is an English habitation name from Gatacre in Shropshire. It comes from 'Gat' – goat; 'acer' a field – an opening in the forest where goats were herded. Sir Bernard Burke from Burke's Landed Gentry said the Gatacres could have been at Gatacre two hundred years before the reign of Edward the Confessor (1042–66).

There was trouble for the Gatacres in 1345 when Geoffrey de Gatacre died and his son Thomas held the estate until he died in 1367. An enemy of the family accused them of not swearing fealty to the King.

Alice, the seventy-two-year-old widow of Thomas, rode on horseback to London and appeared personally before parliament, pointing out that her late husband had fought for the King in Scotland in 1314 and so proved their title to the estates. She was given a standing ovation by the house for her courage, and on 5 June 1368 the estates were returned to her.

The 'Fair Maid of Gatacre'

The Gatacres were famous for the 'Fair Maid of Gatacre', who was a great beauty, in both face and character. She was married twice, one of her husbands being a Wolryche of Dudmaston. She wore a jewel which is still in existence, an amethyst locket, to be worn by the women of Gatacre since, but only while they were single. It was said to be unlucky for a married woman to wear it so when it was worn at her wedding, it was to be taken off by the bride during the marriage service.

The following newspaper article described the sale of the jewel.

A pendant for virgins only

On Wednesday, Sotheby Parke Bernet will be selling at a jewellery auction in New York an unusual pendant, 'The Fair maid of Gatacre', which has been in the Gatacre family of Shropshire since the early

16th century .It has been consigned to the sale by Edward Gatacre of Claverley, near Bridgenorth in Shropshire. The pre-sale estimates range between $15,000 and $25,000 (about £8,380 and £13,960).

The legend behind the pendant is that it was passed down, virgin to virgin, in the Gatacre family for the past four centuries. When a daughter married, she was obliged to hand it to the next maiden sister. It was originally named after Mary, 'The Fair Maid of Gatacre', only daughter of Gatacre of Gatacre, who married John Wolryche in 1528.

The Pendant is regarded as a superb example of Renaissance jewellery. In a frame of gold and enamel, the 16th-century craftsman set a Roman amethyst cameo (late 3rd Century or early 4th Century), carved in high relief with the head of a woman. After recounting the virginto-virgin tradition, an. auction source said quietly to Mandrake's man in New York, 'I don't think anyone dare ask why Edward Gatacre is selling now.' Mandrake, with uncharacteristic audacity, decided to dare. But enquiries in Shropshire, London and New York have revealed that Mr. Gatacre is now out of reach in British Columbia. So Mandrake cannot enlighten the nosey reader as to whether the pendant is being sold on account of a shortage of ready cash or simply a shortage of virgins, or what.

The pendant was bought by the Victoria and Albert Museum in London.

Gatacre Hall

In 1160 AD there was a Sir William de Gatacre and there is a record of his having been an under-tenant in the reigns of Henry II and Richard I. He appears to have been succeeded by Sir Robert de Gatacre who, in 1203, got rights in a moiety of Magna Lya (great Lyth, near Shrewsbury).

A little later, in 1255 Stephen de Gatacre held land at Gatacre and Sutton; he did not attend court, being withheld by infirmity. Sutton was then an appurtenance (an accessory, belonging) of Gatacre.

These gentlemen are sure to have had a substantial house at Gatacre. The Victoria Country History says it was nearly an exact square; at each corner and in the middle of each side and in the centre were

immense oak trees hewn roughly square and without branches, and set upside down so that the roots with a few rafters formed a complete arched roof. The floor was of hewn oak boards three inches thick.

Samuel Bagshaw's History, Gazetteer, and Directory of Shropshire, 1851 says that the ancient house was remarkable for its construction, and has been taken down for some time now. It quotes Camden speaking of this house: 'It was built of a dark grey freestone coated with a green vitrified substance about the thickness of a crown piece. The hall was nearly square and most remarkably constructed.'

Whether this house was built as early as 1200 or earlier we do not know; whether the freestone walls were built round the original house is again conjecture.

Some of this freestone coated with green vitrified glaze is preserved, and a sample of it was sent to the British Museum by Galfrey Gatacre, who was always referred to as 'The Squire' for his observations.

When in England I visited the Hall and brought back some samples of the green freestone that I found in the cellar, which runs the full length of the house. The house was known as 'the glass house'.

The Hall was altered by the addition of wings in 1812 and later more kitchens and servants' quarters were added in 1840.

Until 1916 the Gatacre estate was going along as normal; however, in 1916 Captain Edward Gatacre was wounded in action in France and died of his wounds. Therefore his younger brother Galfrey inherited the estate. As at this time he was farming in Canada, he came to England for a while to get things in order. However, he decided to go back to Canada, leaving the Hall with no-one living in it. Vandals moved in, wrecking the place, stealing books from the library, etc.

There were 800 hectares attached to Gatacre Hall, which were tenanted out to share farmers; this arrangement continued for many years.

Galfrey died on 17 April 1973 and was buried at All Saints Church, Claverley, as are many other Gatacres. The family have been part of this church for at least 900 years, building the north Gatacre Chapel during the 1400s and adding the South Gatacre Chapel between 1500 and 1550. It is under this chapel there is a huge vault in which the Gatacres have been buried right up to 1918. He was succeeded by his eldest son Rex Arnold who died three years later; he was in turn followed by his younger brother David Galfrey, who decided to sell the Hall and 400 hectares to Michael Ryan, a civil engineer who intends to restore the Hall to its former glory. He is restoring all the out-buildings first and selling them off to help pay for some of the restoration. There are forty rooms in the Hall so it is a big task. He has cut down some oak trees on the estate and is using the timber.

There are no Gatacre family members farming in England today. David and Edward Gatacre have ranches in Canada. Edward Victor, referred to as Peter, has a magnificent estate, De Wiersse, which is just out of Vorden in the east of the Netherlands beyond Zutphen. The beautiful home with 70 acres of garden is open during the year. The remainder of the estate is tenanted out to farmers.

All Saints Church Claverley
The Gatacre family has been associated with this church for 900 years, with family members buried in vaults under the South Gatacre Chapel, which the family built, along with the North Chapel. There is also a yew tree said to be 2500 years old in the churchyard.

My father, Reginald Henry Winchcombe Gatacre, was born in 1875 and lived in Bath, England. He attended Bath College and, after finishing school, he passed the entrance exam to the Royal Military College, Sandhurst, to follow his father into the army. The family had been involved with the British army for hundreds of years, including Major General Sir John Gatacre and General Sir William Forbes Gatacre, who has a street named after him in Kalgoorlie. The main streets being named after generals in the Boer War.

However, after receiving a letter from his uncle Charlie, my father decided on a different path. His uncle had started a wine and spirit business in Maryborough, Queensland, which was very successful, and he wrote telling my father all about Queensland. My father then decided to go to Maryborough instead of joining the services at the age of eighteen. He started doing a bit of cane farming at Tinana Creek, near Maryborough, and ran a few cattle as well as doing other jobs.

He was a keen and talented cricketer and played for the Maryborough Past Grammar's cricket team, where he won a few medals. He went back to England on a trip to see his father in 1898 and on his return journey on a boat coming out to Melbourne, he met the woman who was to become my mother. Christian Esson Gordon was from Ellangowan, Banchory, near Aberdeen, and they fell in love. They arrived in Melbourne on the Talune on 14 August 1899 and had decided to marry in five years' time.

In the meantime, Dad purchased some land at Wooroolin in Queensland's South Burnett district and developed it, building a house. Mother stayed with her sister, Elizabeth Gordon, who was now married to Askin Foster; they lived at Boisdale, Maffra near Sale in Gippsland. Born Elizabeth Gordon, she later became Elizabeth Ormond when she and her brother, George, were adopted by Sir Francis Ormond, a great friend of Mother's father in Scotland, before marrying Askin Foster. Sir Francis Ormond was born at Aberdeen Scotland on 23 November 1827 and was one of the nation's greatest benefactors, the first man to establish an Australian Working Men's College and founder of Melbourne University Ormond Hall. He gave most of his immense wealth away to charities.

My mother stayed at Boisdale for a while and helped her sister, then went to England before working as a governess in Ireland and Spain.

She then returned to Australia after the agreed five years to marry my father in Maryborough in 1904. They went to live on the farm at Wooroolin. Another Gatacre, Frank, a cousin, had a farm not far away. There were various members of the family spread throughout the state.

I was born in 1905 and only fifteen months later my brother, Galfrey, was born at Kingaroy, and as children we spent a lot of time together. My sister, Louisa Sophia Elizabeth, was born in 1911 in Toowoomba, where the whole family stayed until she was ready to return home to Wooroolin.

On 20 December 1916 my grandfather died in England, and because it was in the middle of the First World War travel was difficult. So it was not until May 1917 that my father was able to return to England to sort out his father's affairs. He left on the RMS Niagara, arriving in London in early July. He stayed in the UK until September, coming home via America and Canada. He left Vancouver BC for Sydney on the 27 October 1917 on the Metagama.

TOP: Gat's grandfather, William Melmoth Gatacre
LEFT: Gat's father, Reginald Henry Winchombe

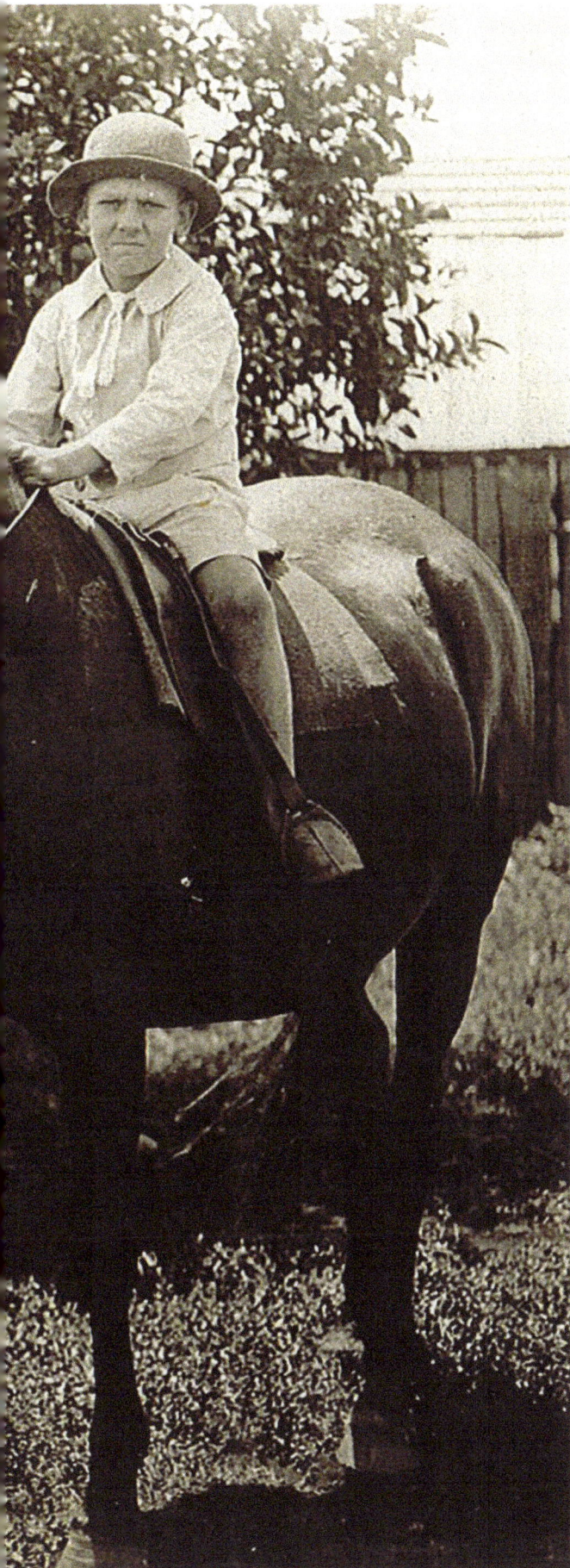

2 MY EARLY LIFE

Gat and his brother Galfrey
on their ponies, 1915

I WAS BORN ON 14 SEPTEMBER 1905 at Nanango, in Queensland's South Burnett district. I was then taken home to our farm at Wooroolin. Mother and Father took me in our buggy pulled by a pair of greys called Mist and Frost. Mist was darker than Frost and together they made a fine pair of buggy horses.

As I recall, our farm was a square mile, 640 acres (nearly 260 hectares). It was rich volcanic soil with some ironbark forest land. I believe it was a forty-inch (102-centimetre) rainfall so the crops grew extremely well, producing good yields. I remember that the disc plough would go down to its full depth and there would be no clods of hard soil. The plough was pulled by four horses. Apart from the forest land there was thick scrub with trees about sixty feet (twenty metres) high. It was very dense with vines growing up and between the trees, plus some other undergrowth, very difficult to walk through when shooting wallabies and pigeons as we had to crawl low like the wallabies. When this scrub was felled by axe, we tried to make all the trees fall the same way. After it was all dry it was burnt on a day when the wind blew in the right direction to get a maximum burn. Then this land was ploughed and crops of maize (corn), oats or lucerne were grown.

When the corn was ripe it was pulled off the stalk by hand and thrown into a dray to be carted by horse-drawn dray with high sides and put under the hay shed. When it was all carted in, a contractor with a corn thresher would arrive. This was a steam engine pulling a machine that threshed the corn from the cobs. It was then bagged ready to be taken to the railway to be sent away. The crop was grown along the boundary next to the railway line so it wasn't far to transport it to the railhead.

We would often grow watermelons along the boundary, when the maize was finished. Someone would pick these watermelons and put them on the posts for the engine drivers. The train came once or twice a week and we made sure any ripe ones were placed there ready for them. If we didn't and the engine drivers saw some in the paddock they would stop the train and go and tap them to see if any were ripe. We could lose quite a few melons if this happened, as the engine drivers made holes in them and might open three or four before finding a ripe one.

OPPOSITE: *Wooroolin School, to which Gat and Galfrey rode their ponies*

We had about twenty-five sheep for killers which were locked up in a high fenced yard. In 1905 farmers were just starting to use the land around Wooroolin for cropping, and some peanuts were planted along with lentils and navy beans. There was also dairy farming and timber milling; prior to this there were large scale pastoral runs with mainly beef cattle.

Life as a young boy

My mother had started teaching me before we had a French governess for a few years before we went to school in Wooroolin. Wooroolin had only a school, general store, butcher shop and a hotel. At ages seven and eight years, Galfrey and I rode our ponies to the state school at Wooroolin, approximately a ten-kilometre (about six-mile) trip. Most children rode their ponies to school unless they were taken in a gig by their mother. The school had either male or female teachers.

I cannot remember playing much sport, except for rounders. I suppose organised sport was difficult because all children had to hurry home to do jobs on the farm. The school was built on posts ten

feet (about three and a half metres) high, which made a nice cool area underneath, where we all had our lunch in the summer. It consisted of two classrooms with a veranda back and front. There were forms to sit on and a wooden rainwater tank which always had cool water in it. I remember one day while eating my lunch a bully provoked me so much that I let go a left punch that fortunately landed on his nose. Then there was a big disturbance as no-one, including the teacher, could stop his nose bleeding. I think it was still bleeding when he rode home. I was not a pugilist but I had learnt from a book called The Boys' Guide on How to Give a Straight Left. This book told how to play all sports, games and many other things, but I happened to remember well the chapter on boxing.

Mother would drive Icey, the old racehorse, in the gig to town once a month to get stores. Icey always had a net over his rump; if not, when swatting flies with his tail he would get it over the reins and he could then bolt. Fortunately, this didn't happen. I can only ever remember seeing one car back in those days. There was occasionally a cattle dealer who would arrive in his old Overland car, whereas most other visitors came by horseback. Sometimes the policeman would ride around and pay a visit but the one person we looked forward to seeing was an agent from Kingaroy. He used to ride out and always bring us a big bag of barley sugar, which thrilled the young! We did have a tennis court, the surface of which was made from crushed white ant mounds mixed with water; it was a good court. Dad taught us to play tennis but our best sport was shanghai shooting as there were plenty of birds. One day when sneaking through the bush we came across a porcupine and it put up its quills. We got such a fright that we ran home as fast as we could, exhausted and bewildered, as we had never seen or heard of one.

On Sundays or wet days, our neighbour, Owen Postle, would take us out shooting in the big, dense, unfallen scrub. He had a shot gun and Galfrey and I would follow behind him. Pigeons were his main aim, or a dingo or wallaby. To find the pigeons he would make their calls, then when one answered we would sneak to where we thought it was, searching the treetops until we spotted it. We got long-tails and bronze-wings. Apart from our shanghai shooting we were also keen bird

nesters and had a lot of pretty eggs. My prize egg was a bowerbird egg which was as big as a pullet's and looked as if red and black lines had been drawn round it. Another egg I collected was a quail's which was red. The most uncommon nest was a speckled warbler bird's, which we got in the dense scrub. It had a pipe entrance about six inches long to the nest, made of spider web, fine leaves and grass.

We had cardboard boxes for the eggs which were divided into two-inch squares and came from the store. They were twelve inches long, three inches wide and one inch deep with a glass top; they were very good. In these we kept our eggs in cotton wool and had big collections of butterflies, moths, beetles and other insects. It being a tropical area, they were plentiful.

Owen Postle also took us to rob beehives. Firstly, he would cut down the tree which had the hive so when the tree fell he could plug up the bee's entrance with a bag. We would then light a fire and smoke it out to keep the bees away. He would cut open the hive, take out the honeycomb and put it into a four-gallon kerosene tin. Mother could then put it into a bag beside the stove and let the honey drop into a basin. It was always very tasty.

We would catch wallabies, bandicoots and possums by setting snares made with snare wire. A snare was a slip knot large enough for the animal to put its head through. Bandicoot snares were put under a wooden gate into a yard containing a barn for grain and hay, etc. Wallaby snares were put on wallaby tracks round the dense scrub. Possum snares were put on about an eight-foot pole leaning against a tree that had a lot of possum scratches on it. Possums like going up a slant rather than climbing straight up a tree from the ground. We thought we would make a lot of money from possum skins: as the possums were twice as big as those down south (like most

Gat aged fifteen

things in Queensland!). On our first night, we caught five but after each skinning about half a possum we gave up as it was a very difficult task and brought to a close the possum-skinning exercise. The only money I ever got was when I shot about six crows and six scrub magpies or currawongs. I cut the heads off and put them in a cardboard box. When the big day arrived and we went into Kingaroy, I took my box to the Shire Clerk, who gave me twelve shillings. He did not see that half the heads were not crows, as they were a bit too smelly to examine carefully!

Every night at dark we could hear the curlews and then the mournful howl of dingoes. Every morning we heard a big chorus of magpies warbling on a big dead tree about a quarter of a mile from home. There were a lot of snakes about too, black snakes, death adders and whip snakes, which often lay along a grape vine or on any branch. The old carpet snake was of course harmless and grew to about ten feet. I remember a few of the big ones loved lying along rafters in our hay shed. They seemed to sleep a lot but they did keep down

the mice! The whip snakes were thin, green and about three feet, six inches (a little over a metre) long. A nearby farmer had one as a pet and I'm not sure if they were poisonous or not. He would let it loose on our billiard table and we often had the gramophone on as well and it seemed to like music and would raise its head toward the gramophone, taking no notice of anyone passing round the table (except one fellow, to whom it showed its fangs).

Mother had a plumpish cook named Ada from Hull in England and she had a broad accent that amused us. She also wanted to ride a horse so one time she had a ride on Galfrey's pony, Ginger. She fell off and after getting to her feet, boxed poor Ginger's ears!! Poor Ginger was blameless. Unfortunately, our pet kangaroo Joe, attacked Ada one day when she was hanging out the washing so we had to give it away. It was a shame as Joe amused us all with things such as liking a cup of tea every morning.

We had a governess for a while who would take us for a walk after school. One day while walking, Galfrey asked her a good question: 'Miss Todd, if God spat from heaven and it hit you, would it kill you?' We wondered about a lot things, such as what made the noise of thunder? We asked our neighbour, Owen Postle, and he told us it was God driving a bullock wagon with tanks on it over big stones. I remember the storms approaching from the Bunya Mountains with lightning and loud thunder.

I think the Postles, who were some of our best friends, were the first to grow peanuts, which today is a big industry in the area. They employed two Aboriginal men and their wives to work the peanuts, which they planted in rows. They used hoes for weeding, a fork for digging and a hand-machine to thresh off the nuts. The men would gather wood of an evening for the women to carry in sacks on their backs while the men walked in front carrying nothing!

One night I remember coming back from the Postles' in the buggy with Mum and Dad, and Galfrey and me sitting in a little front seat. We came across some young curlews and, of course, Galfrey and I wanted to catch them and keep them as pets. We got out and caught the curlews. We were dressed up in our best clothes of course for our outing and after about half an hour the blasted curlews had relieved themselves all over us! We were in such a mess that we chucked the curlews out straight away and that was the end of our pets!

There were five Postle brothers who were all good runners and trained on the farm together. Arthur Postle went on to become a world champion athlete. He gave Dad the medal he won in Johannesburg for the seventy-five-yard record; we had it for years. It was a beautiful medal but I sent it back to his family shortly after Dad died. Arthur went all over the world and when he was in England he married the champion swimmer, Edna Leadbeater, who was the first woman to swim the English Channel.

Gat aged sixteen

Gat about to go shooting

Another memory: Dad thought he would have a few sheep on the place for mutton and as another of our neighbours had been a butcher he thought between the two of them we could all live on mutton. The fifteen to twenty sheep had a little twenty-acre paddock and we used to have to put them in a high fenced yard each night because of the dingoes. Around the bottom was lots of barbed wire and posts to stop the dingoes scratching under to get in. This worked well for a while until one night somehow the gate was busted open and the sheep got out. The next morning they were all dead or half-dead; the dingoes had got them all. That ended our mutton enterprise!

Boarding school

Our boarding house had about twenty acres on the river, where there was a jetty extending about thirty feet, and a bamboo-posted enclosure as a swimming pool, safe from sharks. A master about six foot four tall taught us to swim by pushing us into water well over our heads. Fortunately, it worked and we both managed to swim although we didn't think much of our master, Mr Riding. Here, I also tried fishing for the first time and caught an eel and a stingray. Every morning before school we had prayers in a chapel of the Cathedral. This I will always remember as it was during the Second World War. Regularly we would see hospital trains arrive with wounded who were then taken by cars and ambulances to the Kangaroo Point Hospital.

At school in football practice games, to differentiate between sides, one side rolled their stockings down as, of course, we all had the same jerseys.

We then moved to Brisbane a few years later and lived at Clayfield, where Brisbane Boys College was, so we then attended it instead. It was here that, at thirteen years of age, my brother, Galfrey George Ormond Gatacre, decided to join the Royal Australian Navy (RAN). He passed his entry into the RAN and went to the Naval College at Jervis Bay. Having an influence on his entering the navy was Galfrey's godfather who was Sir Reginald Tupper, a distinguished Admiral in The Royal Navy.

I think 163 boys passed the written examination but after a medical and navy board exam there were eleven left. The Navy Board took note of the boys' appearance, personality, posture, manners and their answers to questions, as well as how they shut the door! Galfrey passed out from Jervis Bay as a midshipman. Galfrey rose to the rank of Rear-Admiral. He was the navy's most decorated officer, including a DSO for his role as navigator on the Rodney when it sank the German battleship Bismarck, while he was on loan to the Royal Navy.

I stayed on at school until I was seventeen, thinking that I may go into the army or if I had enough brains I may go to university. I kept thinking about the country though and I wanted to be a station manager, so eventually my parents reluctantly let me leave school.

Galfrey Gatacre with HRH Prince Philip, London, 1956

3 COOINDAH

A young Gat on horseback

In 1922, I LEFT SCHOOL and went to a place which was owned by an old friend of my father's, Dick Tancred. It was a cattle property within one hundred miles of our own farm. My parents did not wish me to go on the land until I had some professional or trade experience first. I enjoyed my stay with the Tancred family, working under Dick Tancred. I caught the old train to Kingaroy where Dick met me in his old Tin Lizzie (which was a term used back in the early 1900s for all Model T cars) to take me back to 'Cooindah'. Here I learnt to be a cattle stockman and this was because Mr Tancred skilfully imparted his knowledge and helped me to understand cattle.

Gat, the young jackeroo

Cooindah had a twenty- to thirty-inch rainfall and the country was undulating and timbered with big ironbark and stringy bark Eucalypts. I had my Pampa stock knife in a pouch, grey moleskin riding trousers (narrow legs) and well-oiled concertina leggings and of course elastic-sided boots. The leggings and boots were made to look as if they were broken in, as I did not like being a 'new chum'. Dick Tancred had been with cattle in Northern Queensland so I received a good initiation. He had good well-bred horses which had to be ridden the correct way, and they were not spoilt. I thought I could ride but I soon learnt the north Queensland style – 1. Catch your horse by approaching with the bridle over your left arm. Slip on the bridle. 2. Make the horse face you and slip the saddle over the wither and lower gently onto its back. 3. Tighten up the girth and move the horse forward, then re-tighten and put on a crupper. 4. Tighten the near rein, face horse's rump and put left foot into the stirrup, right hand on the pummel and left knee into the shoulder and pull and lower yourself gently into the saddle. 5. Always ride with one's weight in the stirrups, sitting up not down in the saddle (except when walking). 6. Walk the horse until it feels warmed up before trotting or cantering. 7. When trotting and cantering, one always stood up, balancing on the horse's wither so it took a lot of the weight off the horse.

I learnt you must take horses quietly. One day I was reprimanded for not walking my horse far enough to get it used to the girth, and she began to buck. Dick Tancred was watching and dressed me down because the horse was being taught bad habits. I never forgot this and have impressed it on all young jackaroos learning to ride. I spent quite a bit of time riding around mending broken fences. I carried twelve-gauge wire and a wire strainer that worked the same as the old fork strainer. This was two pieces of rod and could be put together making a T-hole in the end of the longer part to put the wire through. It worked well. At paddock corners, there would be a small coil of wire in case one required more. I also learnt to milk two cows once a day and to kill, cut up and salt a beast (beef) as well as a pig. Salting was a cold job in the early morning.

Gates were used in the middle yards and slip rails on the outer yards. Paddocks, too, had slip rails instead of gates and there was plenty of timber available to make the slip rails. (Slip rails are a set of

TOP: *Mrs Tancred, Cooindah, 1922*
ABOVE: *Gat, Cooindah, 1922*

horizontal fence rails that form a temporary barrier in a space instead of a fixed gate. They can be easily removed to leave a gateway.) I lived with the Tancreds as there were no other men on the property. The Tancreds also had a scrub farm about fifteen miles away. The scrub farm was an area that was felled and sown down with Rhodes grass or paspalum. The grass grew quickly after some September storms and cattle fattened quickly, thereby beating the western cattle (which are referred to thus because they graze out in the dryer rainfall parts of western Queensland and hence are slower to fatten) to market by a few months. This was a success as cattle prices were down. I think a fat cow was about two pounds ten shillings.

I helped a man plough twenty-five acres with his two horses and it was said that twenty-five acres of cotton returned five hundred pounds. We camped out while undertaking this work and we sowed it by hand. First, we dropped the fluffy cotton seed into a fire to burn off the fluff, then we soaked it in a tub of flour and water drying it. Then we coupled two slabs of timber six feet apart. Drawn by a horse, it made furrows and dropped the seed in the furrow marks which were about three feet wide. Unfortunately, I left before it was harvested!

Mark Twain visited some large cattle properties during his stay in Australia. One property was Tabinga, a neighbouring property to Cooindah. Here he was shown some camp drafting. Camp drafting is drafting cattle (separating single beasts or pairs or small groups) on horseback and putting the cattle into different mobs when they are out in their paddock. Mr Youngman, who owned Tabinga, wanted some help so Mr Tancred and I rode over to help do some cutting out. He wanted to cut out some fat cattle to send to market and did not want to yard them because they get bruised in the yard. There were about four or five hundred cows in the mob.

A stockman and I held the main mob, while the head stockman, Jack Walters, cut out the fat and older cattle. Dick held the mob of fat cattle and Mr Youngman sat on his horse between the two mobs and wrote down the ages of each beast. Jack called out the year it was born, which was branded on the near cheek of the beast. Everyone was on horseback because the cattle weren't used to seeing anyone on foot, which would have frightened them. This was followed by a scrumptious bush lunch under some big trees, plus a drop of wine to wash it down. When Mark Twain returned to England he said that life in the Australian bush was one continual picnic!! He didn't see the tough side of the Australian bush.

Interestingly, Galfrey was awarded the order of Mark Twain in 1979. He was elected to receive the award from the Mark Twain Journal for his years of Naval Service and succeeded Admiral Nimitz of the U.S. Navy.

Dick Tancred told me a story when we were branding calves one day on a property out west and the owner had a big fellow helping him. A micky (young, wild) bull took to him in the pen and instead of

getting up on the rails he tried to get through the rails and got stuck. The owner who was branding the calves gave him a touch with the hot station brand. There was a court case about it in Brisbane and he was fined about four hundred pounds, which was a lot of money in those days, but he said it was worth the joke!

Another day I rode to the Iron Pot Creek Gymkhana and dance. Tilting the ring was one event; this meant collecting a ring on a three-foot cue at full gallop; the rings were fifty yards apart. Another event was a race where the riders swapped horses and the last horse home won. This was called the donkey race and was good fun because the mounts varied from slow to fast. The old Tin Lizzie vehicle was wonderful as this type of vehicle would go everywhere and had carbide lights. They even crossed dry sandy creeks twenty yards or so wide. One would dig a track about six inches deep for the wheels to travel in and it worked well as digging down you got firmer sand and the vehicle didn't bog or sink.

Another chap and I took sixty cows to Jandowae where they were loaded onto three trucks for market. We dipped them for tick on the way and had a pack-horse to carry our tucker and swag. It was winter and very frosty. As my swag was only a blanket and an oilskin coat, to keep a bit warmer I remember sleeping close to the fire which burnt all night. I think it was the coldest night I have ever experienced. We had to break the ice in the morning just so we could have a wash. It was good fun really; there was no cooking as we had enough salted beef to eat.

I helped erect a single-wire tree telephone line for twenty-odd miles with the assistance of two men, plus two horses and a buckboard with gear and tucker. I think it took us three or four days, but the phone worked.

After six months, I returned home to Father's fruit farm, which he had bought at Raby Bay, part of Moreton Bay, near Cleveland. I had a few ideas of what I wanted to do. One was to get a job at Lake Nash on the Northern Territory border and after a year do a droving trip to Adelaide – five months on the road at five pounds per week which was a very good wage back in those days. Another was to get a job at Thylungra or Darr River Downs which were two large cattle stations.

4 BOMBYANNA

Gat on Emma, Nindigully, 1922

UNFORTUNATELY, NONE OF THOSE IDEAS EVENTUATED so, in 1923, I ended up going to a property called Bombyanna near the small town of Lalaguli, a sheep and cattle property between Goondiwindi and St George owned by the Parker Family. This coincided with a time when prickly pear had taken control. It might be worth taking time here to have a look at the incidence of this pest, and how it was dealt with.

The first recorded introduction of prickly pear was attributed to Governor Phillip at Port Jackson in 1788. It is thought that the shipment comprised drooping tree pear and possibly one or two other species. The reason for introducing the plant was to create a cochineal industry in the new colony. Cochineal is an insect that feeds on certain species of cactus such as prickly pear, and from which scarlet dye is obtained. This dye was used to colour the distinctive red coats of the British soldiers at that time.

Acknowledged as one of the greatest biological invasions of modern times, the introduction and subsequent spread of prickly pear into Queensland and New South Wales had infested millions of hectares of rural land by the 1920s, rendering it useless for agriculture. Prickly pear proved so difficult and costly to control by chemical and mechanical means that enormous areas were simply abandoned by their owners.

Gat with Nigger, Nindigully, 1923

The plant's main dispersal method occurs via the tough, coated seeds that pass undamaged through the digestive system of animals and birds. The heavy crop of fruit produced by pear plants is particularly palatable to birds such as crows, emus and magpies. Many new infestations probably occurred through the germination of seed in bird droppings.

Control methods, such as digging up and burning, and crushing with rollers drawn by horses and bullocks, all proved to be of limited use. However, the introduction of the cactoblastis moth (Cactoblastis cactorum) proved to be spectacularly successful in destroying the weed. Following mass rearing of cactoblastis, millions of eggs were distributed throughout the affected areas of Queensland in 1926 and 1927, and by 1932, the stem-boring cactoblastis larvae had caused the general collapse and destruction of most of the original, thick stands of prickly pear.

The control of prickly pear by the cactoblastis moth is still regarded as the world's most monumental example of successful pest plant repression by biological means.

Returning to my move to Bombyanna I caught the South-west Mail train which arrived about 4 a.m. and was met by the owner's son, Brensley Parker, with a hurricane lamp. Lalaguli had its own siding and there was a path (like a levee bank) from the train to the homestead. We walked across to the barracks, which was a little house not far from the homestead and a couple of hundred yards from the railway line. It was a cottage, four small rooms with a verandah both back and front, where we slept and dressed. There was an overseer and two or three jackaroos. I remember jackaroos had an award wage of twenty-seven and six – first year, thirty-five shillings – second year and full pay of three pounds – third year.

We had our meals at the homestead and did our own washing except for our white shirts which were washed up at the homestead and worn for dinner. Breakfast was at 6 a.m. and we started work at 6.30 but we had Saturday afternoon off, although due to the drought at the time we usually only had the odd day off. During the drought the sheep and cattle had to be fed daily because of the lack of natural pasture.

TOP to BOTTOM: *Poor pear eaters after being fed corn, 1922;*
Bob Absull, Nindigully, 1923;
Gat, Ted and Stan at the fire, 1923

41

The homestead was a couple of hundred yards from the railway and the woolshed was right on the railway line, so sheep or wool could be loaded straight out of the shed and onto the trucks. There was also a dip built high up off the ground where you could run sheep up into the shed and then into the dip. Sheep were treated with arsenic-type powder in a plunge dip, off-shears, to prevent or eradicate the infestation of external parasites, lice or itch mite. It is effected by saturating the wool all over the sheep with an insecticide solution, usually by the use of a plunge dip. The sheep are forced into a narrow bath along which they actually swim and during which time their heads are saturated by ducking them right under.

We were all happy and made our own fun – simple with no hangovers!! I do recall a station hand once getting the train engine driver to bring him six bottles of beer in a sugar bag. He asked me to have a drink with him, but I can't say that I enjoyed warm beer! The country was flat with dead Coolabah trees and red soil. One big paddock of ten thousand acres was covered with prickly pear and brigalow trees, so one could only ride through by following the bullock tracks.

I will always remember my first day of work; we rode out to a paddock and mustered the rams and took them to the yards, which were about a quarter of a mile off the Weir River. It was an extremely hot day and we had the rams mustered by lunchtime. Mr Parker was to arrive after lunch and draft the rams but he didn't come till about 2.00 p.m. I was very thirsty because I didn't have a water bottle with me. I'd ride across to the river every half hour or so and put my quart pot in as deep as I could and try to find a little cold water off a log. By the time we got back to the yards we were thirsty again.

There were four or five of us out mustering and one chap said, 'We better give the new jackaroo a nick name'. A few names were suggested: 'Bronco', but they decided they didn't know if he could ride well enough; and another suggested 'Browser' (a browse is a young sprout) and I have had that nickname ever since.

My next job was to ride the Weir River to pull out any cattle or sheep that had got bogged. There was a drought on at the time and sometimes they walked out too far to get a bite of water hyacinth;

and then they would get bogged trying to get out. We rode the river twice a day, morning and evening. Mr Parker also said if you find any mickies (bull calves that haven't been marked, or de-sexed) throw them if you can, tie them up and mark them. That was right up my alley so I learnt how to throw a beast and tie it up on my own; I knew what to do so it was just a matter of practising and having a good horse, which fortunately I had. Mr Parker said that if any sheep came in that hadn't been shorn to take a chaff bag from out of the shed and a pair of shears and shear the sheep and bring back the wool. I had never seen a sheep being shorn, but when a woolly sheep came in I caught it and tied its legs up with my belt, to lay it on the ground. So I shore the top side first, then rolled it over and shore the other side. I forget how long it took me; it was my first bit of shearing.

It was a dreadful place for goannas, too. There were a lot about and old Marconi, a travelling salesman, used to sell his goanna's salve,

Ted Riley, Nindigully, 1923

a cream that was used for the temporary relief of arthritis pain. In Brisbane, he got a lot of his goannas from this area to make his goanna oil and salve. I remember he used to give a prize each year for the person who sent the biggest goanna.

One day I saw a goanna come out of a poor bogged cow which was still alive! Sheep entrails up the bank were also common. I only had a whip but wished I had a gun, I would have shot the lot of them. Another time I saw a big goanna come out of the water carrying a three-pound black bream in its mouth. One crack of my whip and the goanna was up the tree and the fish was dropped; it was still alive and had no teeth marks on it. I scaled and cleaned the fish and put it in my saddle bag. I took it up to the house where it was greatly appreciated as our dinner that night.

One time we mustered a ten-thousand-acre scrub and pear paddock for cattle. When we had them together the boss rode up on his favourite horse and tried to show us his horse skills in throwing a micky; but he ended up in a heap. He rose quickly because he had fallen on his backside and set off a box of wax matches. His trousers were dropped but he had a blotchy brand for life – not funny but we all laughed.

Once I said to the boss that I had pulled a ram out of the river after it had drowned. He asked, 'Had you fed them?' I replied, 'Yes.' He said, 'Goodness, lost a pint of maize too!' Besides these jobs, I cut and fed corn to the rams which were Wanganella Estate blood line, a stud which I became acquainted with later in life. I remember putting these rams in a yard about half a mile off the river and waiting for the boss to draft them. It was about 120 degrees Fahrenheit so I rode to the river, dipped my quart pot down as deep as possible to get the coolest water. By the time I'd ridden back, though, I was thirsty again. I was like a dog with its tongue hanging out – being Coolabah flood country there was no shade as most of the Coolabah trees were dead. When the river flooded its banks, it spread out for miles. There was a story of an Aboriginal boy caught in the flood who climbed a Coolabah tree to escape, but the water rose higher and when he had climbed as high as he could, all he could do was pray. He said, 'Oh Lord, I am not one of those who is always asking you for help but if you want to do me a good turn, well now is your chance.'

One job which I am glad did not last longer than three or four days was fumigating big rabbit burrows on a sand-ridge. I used a Boston Fumigator which was a small cart with an iron drum, like a twelve-gallon oil drum. A fire was kept in the lower part of the drum and two black bricks in the top were burnt and gave off a poisonous gas. This was pumped down a burrow with a hand-held pump. When smoke came out another hole we put in some bush and covered the hole to stop rabbits escaping. A cranky old bloke we called Baldy Bill was a rabbiter who operated the pump and one or two jackaroos filled in the holes when the smoke appeared. One of his commands amused me: 'Hurry up, there's smoke over there, Jesus, anyone with a glass eye and the sun up his bum would see that!'

I got a bit of praise one day because I found a valuable mare that had been missing. It had been dead for some time so I cut out its brands to identify it and the boss was very impressed by my bringing back this evidence. Probably my worst day at Bombyanna was when I was out one day with Brensley Parker, the son of the owner; there were a lot of wild pigs and it was said that if one rode beside a pig and hit it hard on the backbone behind the shoulders, one could break its back. We came across a pig, so we both got waddies (sticks) and took after it; we belted it with no result. The pig knocked up, so Brensley got off his horse to, he thought, finish off the pig. He gave it a good belt but the pig turned and had Brensley hopping from one leg and then the other to avoid its tusks. I was about to get off and help him when the pig took off and we called off the chase. I am yet to be convinced that one can break a pig's back this way! When we found a sow with young ones though, we would catch a couple and take them back to put in the fowl yard at the homestead, there they could be fed with the chooks.

At Bombyanna, one would sometimes find a sheep walking about with one eye injured. A crow was often noticed on an animal's head, pecking at its eye while the sheep was moving and trying to shake it off. In one paddock, there was a small clump of green trees where there could have been up to fifty crows nesting at night. Mr Parker had four guns so four of us went to the clump of trees one dark night and let off shots into the tree tops, hoping to get the offending crow! Birds flew out and then returned, so we fired quite a lot of shots, but it was too dark to see if we hit anything.

Bob Absull the cook, stock route,
Nindigully near St George,
Queensland

The head stockman, Ted Riley, a jackaroo named Bob Absull who was the cook, myself as shepherd and horses were sent away with four thousand ewes to some agistment near Nindigully, which was a Cobb & Co. Mail Change Depot. It was a stock route, a mile wide and scrubby, and we often had to get off our horses to look over the sheep. Bob drove a wagonette with our tucker on it and he had to go ahead and make a break yard (which is a temporary yard to hold sheep overnight) or repair an old one to put the ewes in at night. The hop bush to make our break was yarran sandalwood hop bush. It was just stacked up to make a fence and could easily fall into disrepair. Sometimes Bob had to do a bit of axe work to make it secure, then he'd prepare our tea. When we arrived, we put the sheep in the break and tied the dogs up in the opening of the break, to hold the sheep there for the night.

Bob proved to be a better axe-man than cook. He made a good stew with three essentials – meat, spuds and onions. His damper was usually a sod (very hard), and one nearly killed me! It was one cold night when I had several pieces (still warm) with treacle, to warm me up. I ended up spending most of the night in pain with indigestion, hanging over the fire for relief. The next day Bob was told by my boss to 'throw that bloody sod away and don't throw it on the track in case a car comes along, because it may cause an accident!' We slept just on a ground sheet; I had a canvas swag then and the next morning we were up at daylight. I had to go and get the horses which were hobbled out the night before and bring them back to the camp for breakfast just after daylight.

Along the track, we came to a selector's place and Ted, my boss, said, 'Slip in and ask that cocky if we can water our horses at his bore drain' (about one hundred yards inside). I went and asked the owner, a big six-foot, six-inch man, who inquired, 'How many horses?' 'Five', I said, so he asked if I had a whip. When I replied yes, the owner told me we could, but if any horse started eating couch grass along the drain, to give him a chop on the nose. He proved to be a good friend because when it rained all night he let us camp under his cart shed. He explained to me that his father was an Irishman and his mother a Scot so, he exclaimed, 'I am a crossbred but by hell I can fight!' We decided not to test him out!! When we got to our agistment paddock we pitched our tent and Bob, the cook, went back to the station. Ted and I mustered up the sheep out of a very scrubby, prickly pear area out into the open and then cut scrub for them to eat.

If they stayed in with the pear they used to get a lot of thorns in their lips which then would swell up so they couldn't eat grass. We would

Jean Parker, Lalaguli, 1923

muster them up daily but some sheep would wander back to the pear area. We would then have to keep those sheep near our camp. We called them the hospital lot as they had sore mouths and we fed them on grain, until we ran out. Kurrajong bushes were not to be cut as they were good sheep feed. But because it was so easy to do, we cut all the branches but one, so the tree would revive. I think we would have stayed there a couple of months altogether. After we arrived though Brensley Parker came and took Bob back. So just the two of us stayed on and looked after the sheep.

After a while we were getting a bit sick of only our damper, treacle and salted mutton, so we rode into Nindigully for a change of menu and some more stores. That night Ted and I had a big night out: we rode to Nindigully for tea and a few beers and after an enjoyable evening, we stayed the night and rode back to camp in the morning. I remember there were a few men shut in their rooms making odd noises. My boss said they probably had the DTs (delirium tremens) as the publican was supposed to make his own beer from prickly pear fruit, because there was nothing else about!! Of course, in those days a man would be on a station for months – often until his skin was starting to crack – then take a breath and go into the local pub. The first thing he'd do was hand his over cheque and tell the publican to tell him when it was cut out. Nindigully happened to be the last Cobb & Co. coach change in Queensland, so they said. I know I saw the old Cobb & Co. yard there where they changed their horses just the other side of the pub. The pub was all that was at Nindigully, no store, etc., and there is still only the pub there today.

There used to be thousands of acres of pear until the cactoblastis moth (Cactoblastis cactorum) was introduced (see above); the caterpillar ate the pear fruit and killed the plants, which fell to the ground like paper. It is now beautiful farming land. I remember the story of someone overhearing some talk about the cactoblastis and how it had cured the prickly pear. Apparently, they hadn't heard correctly and thought it was some Catholic Bastard that had dealt with the pear! Returning, I remember having my coldest shower. It was at a place called the Rockwells, three wells with fresh water on an iron-stone ridge, supposedly made by the Aborigines. The water

TOP: *Nindigully pub*
ABOVE: *Jean Parker and Mel Foote, Bombyanna, Lalaguli, 1923*
OPPOSITE: *Stan Willis cooking a damper, 1923*

was not much below ground level. The wells had a chock and rail fence – logs crossed over about three feet high – round them to keep out animals. It was a cold starry night but we hadn't bathed for a long time so after tea, with a petrol four-gallon tin, we poured water over each other. We did all feel better and cleaner afterwards and soon warmed up at our camp fire.

After a few months, we mustered our sheep and drove them back to a big scrub and pear paddock at Bombyanna, which now had a bore drain. The artesian drain came through a wild dog-proof fence which branched into a dam that they had made and were able to fill up with the artesian water. The drain was ploughed only with mouldboard plough (turns over the soil to one side) and not cleaned out so another chap and I had to shovel out the dirt in about one hundred and twenty degree temperatures. I think it was only a few hundred yards but seemed like a mile, a few days' hard work, I recall. We did get extra help from the homestead too but I don't recommend to anyone, making a drain with a shovel. Another day at daybreak we started calf branding; we scruffed (where each calf is manually caught and thrown down so various procedures can be administered) about three hundred calves after drafting them. There were four scruffers and this was still not as heavy work as shovelling hard clods out of a drain.

The Parkers had one daughter who left school at the end of the year. When Jean was home for mid-winter (three weeks) and (one week) spring holidays, she always had a girlfriend staying with her, so the other jackaroos and I would often stay on after the evening meal and play the gramophone and dance on the verandah. I enjoyed this because Jean was an attractive bright girl, as were her friends. We all went over to Kyawanna one night for a party; it was a good party and later Bill Rae, whose father owned Kyawanna, went down to jackaroo at Wanganella and I met him again. I left Bombyanna to be home in time for Christmas. My boss offered me a rise to two pounds ten a week if I returned. It was nearly full pay (three pounds), but I had made up my mind to join a company out in western Queensland. I was given a short but good reference: 'This lad has ability, strength and character and I recommend him.'

5 BOONOKE

Boonoke special stud rams
for Sydney, 1930

By now it was 1924, and my cousin Phyl Ormond from Melbourne, came up to stay. One of her boyfriends was a Claude Weeks, who was secretary and amateur rider for Mr O. R. Falkiner from Boonoke. Claude arranged a job for me at Boonoke, which was to prove a long and happy association with F. S. Falkiner & Co.

Old Frank Sadleir Falkiner came out to Australia in 1850 from Ireland with sixteen pounds and finished up owning vast land holdings. The Falkiners were in the wool trade in England, and when there was a depression in the trade there they moved to Ireland and continued. Wool was obviously in their blood. The Falkiners were modest, likeable employers and never neglected their employees. Everyone liked them and felt part of the firm. They were generous to the district as well, with few people knowing about their actions.

Phyl Pole and Claude Weeks, 1923
RIGHT: *Gat on SS* Cooma, *Brisbane to Sydney, then onto Boonoke by train*

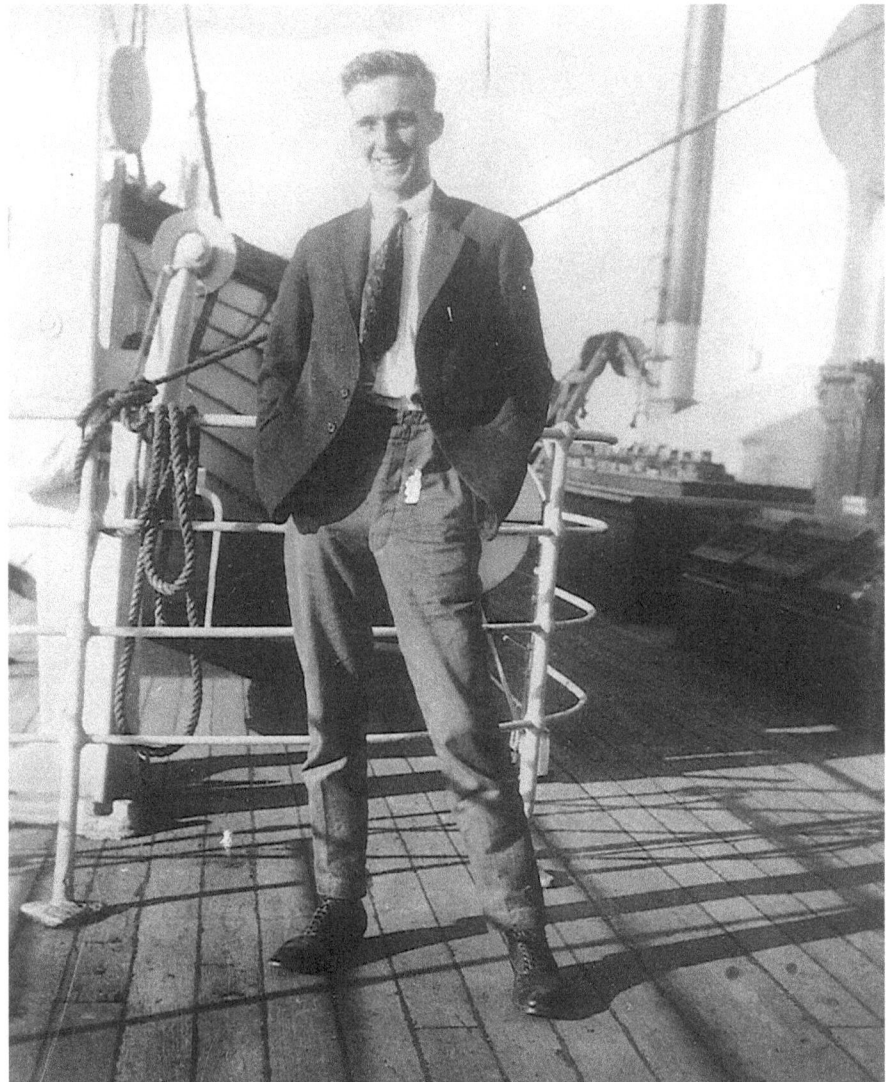

It was interesting that, before I made the journey to Boonoke, my cousin Phyl Ormond and her Aunt Sylvia and husband Alec, with another couple and Keith Murdoch of newspaper fame, all had lunch at Lennon's Hotel in Brisbane. (Some fifty-four years later Murdoch's son Rupert would purchase all the Falkiner land holdings.)

On the day of the lunch, cousin Phyl and I travelled to Sydney together by coastal boat. We arrived at the wharf around 8 a.m., plus Taffy, my dog. We hired a 'hansom' cab (only hansom and double seaters were available for coastal boats). Hansom cabs were drawn by a single horse as the 'modern day' taxi. We deposited Taffy at Central Railway station and put in the day at the Hotel Sydney, near the station. Then after dinner that night I took Phyl back to the boat and she went on to Melbourne while I caught the South-west Mail train at 10 p.m. to Jerilderie. It had been an adventure for me, but then life was simple in the twenties.

Steam engine and pump, woolshed irrigation

On 1 April, I arrived with my bitch, Taffy, at Jerilderie. From there I went to Conargo by mail car and to Boonoke by the station mail gig. I had heard what Boonoke was like and I was not disappointed. I arrived at the barracks that evening and settled in while Taffy curled up under a Pepper tree. I had brought my jackaroo's outfit with me, which included the following: 1 swag; canvas sheet with calico sides; one smith oilskin; one hat; 3 shirts; one white shirt; two pairs riding moleskin trousers; 1 good pair gabardine trousers; 6 pairs socks; two pairs riding boots; two pairs pyjamas; one blanket; 2 towels; one water bag; one pair concertina leggings; one cane-handle whip.

The Boonoke stud was run on the two properties: Boonoke which was 160,000 acres, and it joined Moonbria which was 70,000 acres. On the south side of the Billabong was Wanganella Estate, which was about 30,000 acres and this country ran the Wanganella Estate flock.

The first mention of the area known as Boonoke appeared in Campbell's 'Squatting on crown lands in New South Wales', which appeared in The Government Gazette on the 19 February 1840. During this period the Carne Brothers owned the area. The next recorded owner of Boonoke was Benjamin Holmes in 1858. He then sold it to Lyan McKenzie and Co. in 1859, who then sold it to Butchart and

Boonoke woolshed, 1924

Dunn, who in turn sold to Dutton Simpson and Darlot, who the same year sold it to the Patterson Brothers. They ran cattle on the property before switching to sheep on hearing the success of their neighbours, the Peppins of Wanganella. While the Pattersons had Boonoke, there was a fire at the woolshed in January 1869. It destroyed thirty bales of wool. Patterson later sold out to his neighbour George Peppin and sons in 1873. The Peppins held Boonoke until 1878, when it was sold to Falkiner, Ross and McKenzie, who in 1882 put it up for auction, and Falkiner bought out his partners.

In 1888 there was a shearers strike and it is believed the organiser of the striking shearers set the Boonoke Homestead on fire on the 16 August. The house was almost destroyed with the only area left standing being the store, kitchen and barracks. There was further disaster when Falkiner's own loyal shearers came to help save the place. They found the cellar, got drunk and started fighting amongst themselves, breaking all the furniture. Fortunately the dining room table was saved and it is in the dining room at Boonoke today. The house was rebuilt within two years and it still stands to the present day. However, in 1990 extensive renovations were carried out, virtually doubling the size and complementing the original building.

Management and staff

Boonoke was known as a township without a pub because of its size, buildings and number of employees. There were several employees because the station needed to be able to do any jobs and repairs on the property. The staff included a manager and his wife, with a cook, a housemaid and a gardener. Then there was an overseer, a bookkeeper, four jackaroos, barracks' cook and housekeeper, two windmill men, blacksmith, groom, two boundary riders, milkman/butcher, mechanic and lighting plant man and two teams of horses. There were also three outstation families.

In the barracks were the overseer, the bookkeeper and four jackaroos. Jack, another man and I had beds on the open veranda but dressed inside. There were three bedrooms, a kitchen, dining room, sitting room, bathroom with chip heater and maids' room. We had a cook

TOP: *George Lang loading sheaves of hay, 1925*
LEFT: *Boonoke manager, Joe Thompson*

Jimmy, the veggie gardener
OPPOSITE: *Boonoke horse team with a load of wool*

and housemaid to care for us, i.e. do our washing and mending (better than Queensland). Both the homestead and barracks had EL & E refrigerators, worked from a DC plant (direct current power: substantially steady in one direction in a circuit). Other cottages housed shearers, rouseabouts and boss of the board and staff, and they had kerosene lamps and coolgardie coolers. The coolers came in a hessian container 6-foot-high, 1/3 by 3 feet wide; the breeze blew through the damp hessian sides and the air became cool.

The manager Joe Thomson and his wife lived in the homestead. The homestead was a large brick building situated on the Billabong Creek. The office was at the end of one wing and there were six spare bedrooms for visitors. Also in those days buyers who came to inspect sheep would stay there for a day or two. The overseer was Mr Shallberg when I arrived; however, during the year he was moved to Wanganella Estate as manager and Bertie Howell took over as Boonoke overseer and remained overseer until 1929.

They had a gardener who looked after a large garden at the front and lived in the hut. There was also a Chinese gardener called Jimmy who did the vegetable garden at the back and pumped the water. The water supply pump was at the end of the garden on the creek, near a little cottage, where Jimmy lived and cooked Chinese food for himself. One morning when I walked down to check on the vegetable garden there were four crows on a line with breasts plucked and open. I asked what they were for and Jimmy said if you take out the gullet before 10 o'clock and put it with some whisky, it makes good medicine.

The men lived in the hut where they were looked after by the station cook. There were eight rooms, with one or two men to a room, two dining rooms and a kitchen. This had an open fire and baker's oven for making bread. There was also a pantry and cook's room with a shower and toilet in a separate building. Then there was another cottage on the Billabong where the horse driver or teamster lived. The blacksmith and his family lived in another cottage and their son, a handyman, was in another. There was another cottage on the Billabong in the woolshed paddock. This was occupied by the person who did most of the shed work such as looking after sale rams, feeding them at the shed and during joining he did the hand-serving of all the special ewes (which is the special selection of rams to the selected ewe or ewes). He could fill the Boonoke shed himself with his two dogs; the Boonoke shed had twenty catching pens, and could hold 3500 sheep under cover.

There were big stables, which were brick; on the top there was a loft, where the chaff and oats were kept. From the top, a chute went down to fill a feed tin bag. In the stables, there was also a saddle room, two loose boxes, six stalls and a stallion box stall which opened out to the yard. There was a big set of horse-yards, three big yards and two smaller ones, all with feeders, and there was a buggy shed. A stallion was always kept at the stables and the groom looked after it as well as running the mail three times a week.

To provide fresh meat there was a brick butcher's shop, which would hold a beast for beef as well as sheep carcasses. The butcher killed the sheep to feed the forty-odd employees there at the time. There was also a separator for milk. The butcher used to bring the cows in every evening to take the calves from them and then to milk the cows each morning. The hut cook would ring the big bell at 6.30 a.m. every day for breakfast, 7.00 a.m. orders were taken, 12.00 for lunch

and 1.00 p.m. to recommence work. Men going out on their own would take their lunch with them. This was usually a leg of mutton, loaf of bread, tomato sauce, tea and sugar and a 4-gallon petrol tin to use as a billy. Water bags were carried as well.

Every Saturday the bookkeeper would issue the stores at the homestead for the next week. The groom would take a draught horse and a dray up to the store and cart the stores round to the various places – barracks, homestead and the cottages and hut. If Cocky, the draught horse, was still carting stores around and the hut bell went at 12 o'clock, he would move away himself because then it was feed-up time. He wanted to be unyoked, put in his stall and fed.

Most work was done by two horse teams and two bullocks with wagons. These teams used to cart wool during shearing. After shearing they had jobs such as carting wood into the homestead and shearing shed and making firebreaks. They also cleaned out dams. Spring-carts, drays and gigs drawn by horses were used. The manager had a Ford ute and Austin car while the overseer, a cut-down Napier car or a Ford ute. There were three out-stations, one was at the back of Boonoke near Willurah Station. There was one at the very back called Dormy and the other was the Black Sand Hill. They were all manned by married families. Each place really worked about forty thousand acres. All these places were connected to the homestead by telephone, including Wanganella Estate, Boonoke North and Warriston.

OPPOSITE: *Boonoke woolshed, 1931 flood*
BELOW: *Bullock team, Boonoke*

Party line

As there was no wireless and we rarely saw a paper, the men at the out-stations often used to listen in when they heard a ring on the telephone, just to get a bit of news. One night Mr Thompson was talking to Mr Falkiner at Boonoke North and Mr Carse was at Moonbria. They were trying to think of the name of a ram they sold to some stud, when one of the fellows who had been listening in and couldn't resist it any longer, blurted out 'Bullawarra, Mr Falkiner'. The reply was 'Get off the bloody phone, you old bastard!!'

I was only at Boonoke a day when I was sent out to help the back-station man riding lambing paddocks. A Mr Les McMaster and his wife were the people in charge. Mr Mac was a man with experience, ability and character and his grown-up son, George, was Wanganella Estate's overseer. Before taking over the Conargo Hotel, Les McMaster was shed stud sheep man and known for his ability and his dogs. He had a big black dog, Ringer, who would jump up at the horse's nose. Old Mac used to say 'I'll give it to you, I will, I will, I will' but he never hit him. My camp at the back-station was in the feed shed but I was happy with the Macs. Old Mr Mac drove a buckboard buggy and I rode round the sheep. We used to do three paddocks each day, covering half the paddocks in the morning and the other half after lunch.

Boonoke shed hands, 1924

Homestead men worked on Boonoke, because main jobs were done around the homestead. Each morning, breakfast was at 6.30 a.m. and the homestead employees got orders, i.e. jobs for that day, at 7 a.m. from the stables. The outstation was phoned each morning at 6.30 a.m. or the night before, as all places were connected by a private phone. My job, apart from seeing the stud sheep were well, entailed attending to fences, windmills, water troughs and dams. Oiling windmills I did not like much because I didn't like heights.

Sheep were only yarded when necessary, otherwise they were attended to in the paddock for fly strike and grass-seed. When jackaroos had their horse saddled up, you had your lunch in your saddle bag and quart pot in one side. On the other side, you had your dip carrier and pair of shears and butcher's knife in the middle. A bottle of dip was carried on each side in case you saw a fly-blown sheep while riding around the paddocks. You were expected to look out for any sheep who were wool blind or had seeds in their eyes and to observe how the sheep were doing while you were riding around. If you had to get off your horse and catch a sheep in the middle of a paddock, where there was nowhere to tie your horse up, you tied the rein to your horse's front leg. I generally carried straps though and put a strap round the horse's front legs above the knees and pulled them tight together. The other strap was tied from the girth to that strap so it wouldn't slip down any further. Then I had my reins unbuckled loose on the ground. You also taught your horse to stand without being strapped with the reins loose on the ground.

After lamb marking I was moved back to the homestead for shearing (see below). I was pleased because a jackaroo got one pound per week (first year) but Mr Thompson, the manager, gave shed wages during shearing. Jackaroos were really apprentice stockmen and worked to gain experience. They were looked after but started on little pay, but with experience their pay increased and they could then become senior jackaroo, overseer, then station manager if they were lucky. Station hands got the award wage.

Wanganella Estate had had a beautiful homestead which was burnt down in 1935. It was two storeys with a lovely staircase and furniture, and was regarded as one of the finest station homes in the Riverina, a magnificent Tudor mansion. The day of the fire, a Saturday afternoon, a cricket match was being played between Wanganella Estate and Conargo township. After first thinking it was someone cooking, the players soon realised the house was on fire and luckily, due to some quick work most of the furniture was saved. However, the homestead was destroyed. A message was sent to Deniliquin by phone and a fire engine and pump was sent out; however, it was too late to save the home.

The fire was started accidentally, by the boy who was in charge of the wood-fired water heater, known as a donkey. He overfilled it and had not been able to close the door because he had put too much wood in.

After the destruction of the main homestead the extensive brick workmen's quarters were converted into the main homestead, and this rambling building still serves that purpose today. The original brick woolshed (known to generations of shearers as the 'brick oven') remains, however, and was the scene of a painting in 1921 by the famous war-time artist George Lambert. Entitled 'Weighing The Fleece', it's now housed in the Australian National Gallery in Canberra.

*Pennyfather Snr selecting Zara
rams for South Africa in 1927;
Jim Moore and John Huggan
holding rams*

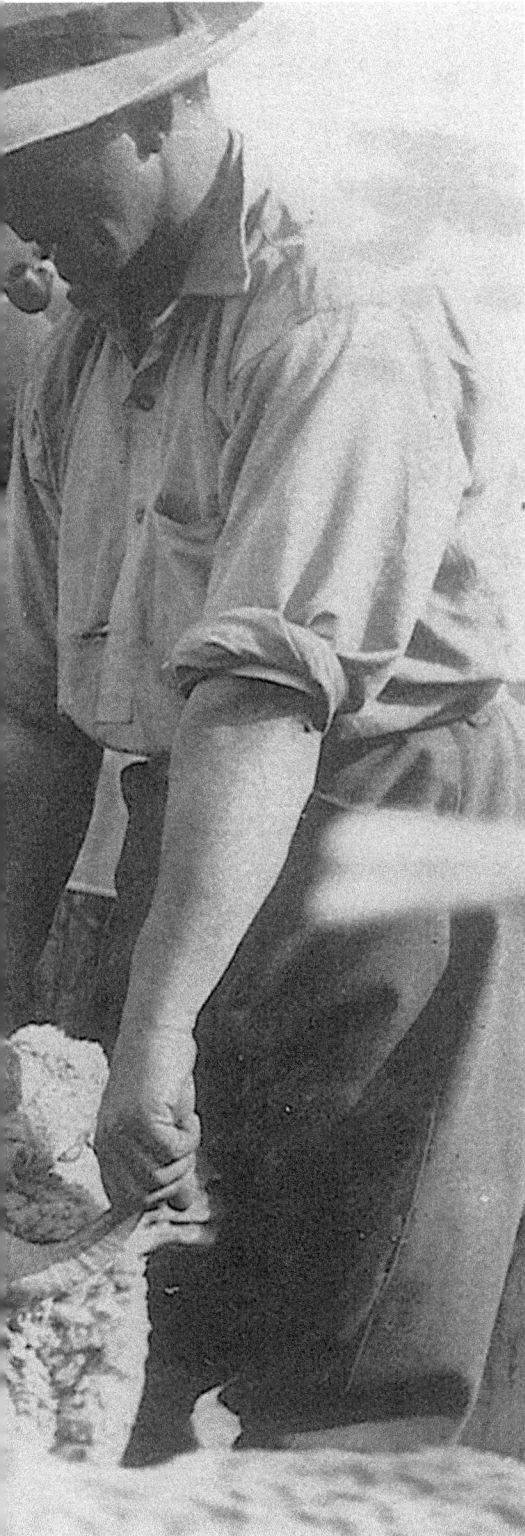

Managing stud sheep

The management of a big stud was different to the flock management. A stud property also costs more to run than a flock place because the sheep cannot be replaced and need more room. Because of the many years and cost of breeding stud sheep they cannot afford to lose any, whereas the flock sheep can be sold if feed is scarce. A stud sheep needs four acres and better feed while a flock of sheep needs only two acres per sheep. The two main aims in breeding were classing and mating to introduce a good, hardy, profitable sheep, which would stand up to hard conditions, light carrying country and long travelling distances to water. Therefore, the growing out or developing of the lambs and weaners was a skilled job that needed good sheep management. This experience is a big help to any flock management. Some jackaroos who didn't come from family properties often became sheep classers and managers, etc., as those with family properties usually went back to their families. Merino sheep were bred to cope with all kinds of varying country, particularly western Queensland. They were noted for their constitution and their hardiness as well as producing a lot of wool.

I remember looking up my geography book to see what the Riverina was like. It was likened to Southern Siberia where they got extremes of heat and cold. This produced hardy animals that survived in difficult terrains. The Riverina grows a bigger variety of grasses, saltbushes and herbage than any other part of Australia. This suits the merino sheep who likes a variety of short feed. They like an area to walk in and look for fresh feed. The area has a twelve to fifteen-inch rainfall. The May–June rains, with some in the spring, will usually carry stock through to the next May. These are the reasons the Riverina is noted for stud sheep breeding.

Shearing

From 1920 the Falkiners did their own shearing, they didn't employ a contractor. This was because one year they had some sheep in the shed for crutching and the crutchers wanted more pay. Mr Thompson said, 'No, the award is the award and that's what you will get.' They considered it and they didn't shear but just left the sheep out in the yard. That year they weren't crutched as there wasn't time; they jetted (see below) for the fly strike what sheep they could before they lambed. Since then no sheep were crutched because we cleaned them up in the yards with shears. From then on, the station did their shearing themselves.

Shearing commenced approximately 18 July each year and included Wanganella Estate and Moonbria sheep. This lasted six to eight weeks and in that time we shore 80,000 to 90,000 sheep. The biggest shearing I saw was 111,000. That was when Mr Ralph Falkiner had a place called Groongol in the Hay area. He sold it and brought down 11,000 wethers and had them on Boonoke. We shore them at the end of the stud shearing, which brought it up to 111,000. He also brought all the Groongol horses down as well and two or three hundred head of beautiful shorthorn cows and calves. These we grazed here for a period and then sold.

Boonoke shed hands, 1925

I spent most of my time at the shed as time was costly and there were 3500 to 5000 sheep shorn each day. The same number had to be shedded up for the next day's shearing also. The yards were all bricked, they were always swept with a hard broom every morning, so the sheep started off on a clean surface. When filling the shed, the yards were made so one could draft sheep up the two back races, which were wide, to fill the twenty catching pens, or under the board and wool room when shedding up. The shed would hold 900 under the wool room, 300–400 under the board and 2000 in the shed. If there were ewes and lambs of course more could be held, sometimes over four thousand.

There used to be ten blade men and thirty machine men in the shed, two men penning up, four wool pressers, thirty rouseabouts, broom boys and table men. Then there was the wool classer, boss of the board, expert and the bookkeeper. The shed hands were engaged by the classer and the manager organised the shearers and supplied the shears. Written applications started arriving by April, as April/May was when applications for the Boonoke shed were taken. It was the biggest shed around the Riverina, having forty shearers. There was no better shed anywhere, it was noted for its excellent shearing team.

A senior jackaroo would usually get the job as shed overseer if he was suitable and able. Two jackaroos would help at the shed and then get full shed wages, so this was a plus for being experienced. There

TOP: *Boonoke blade men sharpening their shears, 1925*
ABOVE: *Shearing, Boonoke shed*

Gat in the Boonoke yards

were two men penning up, too, and I remember one chap was there every year and taught an offsider. A shed overseer told me he asked a shed hand how he was getting along, when he returned one year. The fellow said, 'Well, I am married and got my own house filled with white rose furniture.' In those days, kerosene came in good pine wood boxes and on the cases a white rose was printed (White Rose was an American company that made kerosene).

When shearing began, shearers arrived in gigs and light spring-carts. Three or four station hands would shear too and then maybe go and do a few more sheds before returning to the station for work. Tasmanian blade men would often come over before shearing and do station work until the general shearing. Having the station's own men meant they always had a good team with no trouble. Knowing them personally also enabled Mr Thompson to get rid of any agitators or crook shearers. One Deniliquin shearer who was in the team, we later found out, had been a disturber in the strike. Mr Thompson was walking down the street some time later when this chap came up to him and said, 'How about a pen this year, Mr Thompson?' Mr Thompson, still walking, said 'No'. The chap then said, 'I didn't get a pen; did I get a life sentence?' He replied 'Yes' and then walked on. I know of no better man with Mr Thompson's ability, character and knowledge. I still remember him saying 'Browser, there are no degrees of honesty – either honest or dishonest!' The forty shearers had their own cook. Some wily souls would contrive to get the mailman to bring bottles of alcohol out from Deniliquin. These were then hidden in a sugar bag and anchored in the Billabong Creek for cool storage.

The blade shearers generally shore all the rams and the ram lambs. The boss of the board would count out from the pens. A particular shearer had to leave one year and was replaced by a Queenslander who was supposed to be a real gun shearer. The shearers were averaging about 115 sheep per day and this chap did 115 on Thursday and Friday. The old shearers began talking and reckoned he was no gun but then came Saturday morning and he did 115 again only taking half the time, so obviously could've shorn more on the full days.

The station bookkeeper recorded all wages each week and within two hours of cut out, all were paid. A few unfortunately had nothing left because of playing 'two up'. On Saturday afternoons, the Conargo storekeeper would bring out a load of stores for the shed hands. He had a general store and could supply a pair of trousers or anything that was required. When selling anything, he would say, 'Good strength and it's dirt cheap!'

Bridge being erected over Browns Creek to get sheep to the woolshed, 1931

We got about 3000 bales of wool each year at shearing. Two wool presses were used by four men. The wool room held 10,00 bales of wool and it was full up at the time of the 1931 floods. That was a tough year; we had to cart the wool in small lots and get it out to the main road to have it sent away. One jackaroo would write down the description of each bale in a wool book and two others would roll it out to the platform and load with the help of the teamster. He would also get the job of checking the bales onto the wagon. He would have to check that they were all branded correctly too. At times, he was helped by a shed hand.

Boonoke woolclasser, Bill Taylor, with fleece, 1925

Bill Taylor and the shed expert
at Boonoke woolshed, 1925

Due to the floods we even had to build bridges over some of the gullies to get the sheep into the shed because it was so wet. The paddocks had wild oats three feet six high there and the sheep used to make a track in it like a street and just stay in that line. There was terrific feed that year. I remember it was up to the roof of your car if you drove through it. It was unbelievable really.

Generally wood was cut by swaggies for food for their tucker bag. The shed hands, boss of the board, expert and wool classer had their food and cook supplied by the station. Jackaroos who were at the shed could have their lunch at the cottage. We had a character of a wool classer called Bill Taylor. One tale he told really puzzled me. I was having lunch at the shed cottage one day during shearing, and the conversation got onto marriages. Bill said, 'Well I used to get on well with my wife but my mother-in-law, I damn nearly killed her one day. I attacked her with a picket and hit her on the head.' When shearing extended to five weeks they would be wondering when shearing would finish. The expert, a short, busy little bloke kept on asking because he had to go to Nangunyah. It annoyed us and old Bill said, 'Shut up, you are like a mosquito – Nangunyah, Nangunyah, Nangunyah!'

Slight mishap – shearers on their way to the shed, Boonoke, 1931

Horses were used by all, so if one wished to go to Conargo, Deniliquin, etc., they rode or drove a horse in a gig. Roads were just tracks formed in places, there was no bitumen. There was sometimes a twenty-four-mile ride out to muster a paddock, so a twenty-mile ride to Deniliquin was nothing! At the beginning of shearing though, a description of all horses and owner's names was taken and the horses sent out to the back paddock until the shearing was completed. In an emergency, an employee was driven into Deniliquin in the station ute. During shearing, we worked five and a half days and then Saturday afternoon was for washing, etc. There was no grog allowed on Boonoke so the shearers' only entertainment was generally a big 'two up' game on Saturday afternoon.

When branding, the sheep were run through an arsenic bluestone foot bath. We had double branding races so there was no hold up when branding. There were two horse teams, plus the bullock team. It was hard work being a teamster, but the company was good; on most jobs two, sometimes three teams worked together. While he was working at Boonoke Station there were five other teamsters working there. There was Matt O'Donnell with twenty-two bullocks and Bill Wallace with a horse team of sixteen. Also, Alf Treasure, who had worked with bullocks, but at that time was working with horses, and a man named Tibbens, who came from Gunbower, and Harry Heath. Harry worked for many years on the station and had twenty-six bullocks while I was there.

Harry Heath

I knew Harry Heath and remember him as a reserved and kind man, one who was kind to animals and humans alike. He was quite a character and everyone liked him; they used to call him 'Old Harry'. He loved his horse, dog and bullocks. He rarely used a whip on his bullocks, but when needed, he cracked it close to them, not on them. He never raised his voice, always speaking quietly to his bullocks. Harry was always afraid of lightning and thunder. No matter where he was working or what he was doing, if a storm came, he unyoked his bullocks and either rode home, or got into his hammock which was strung at the back of his wagon. One Sunday I remember particularly well; everyone was out doing their weekly washing, except Harry, who was watching with extreme interest, the groom separating the milk in the dairy. Harry could not take his eyes off the separator, and when the job was done, he thought for a while, then said; 'Well the bloke who invented that machine was no curlew.'(a person referred to as a curlew was someone not having many brains, or an idiot).

Once Harry was cleaning the mud out of a large ground tank with a big mud scoop. His offsider, another old bullock-driver, was helping him by making a team of six bullocks pull the empty scoop back into the tank. Then Harry, on the opposite side would pull the scoop out, full of mud. The offsider then pulled the scoop, which had been emptied, back into the tank. I arrived one day to see what progress had been made, only to find Harry and his offsider, Tommy Gordon, one on each side of the tank, arguing, each man telling the other what he thought of him. The funny part about it all was that each man kept on his own side of the tank, and when each had said his piece, both started work again as if the argument had not even existed.

On another day, I called on Harry when he was making a firebreak between Moonbria Station and Boonoke. It was 1931, a flood year, and there was a terrific amount of feed around. I asked Harry how he was making out and told him that he seemed to be making a good job of it. Harry replied: 'You know Mr Gatacre, I've been up and down, up and down, until I've jolly well-worn the very grass off.' One February, in the early thirties, when the temperature had been

Shearers, Boonoke shed

71

over one hundred degrees for a week, and the first remark in the morning was, 'Well, it is going to be another hot one today,' Harry called at the shearers' hut for a billy of hot water. The Robertson family was living there at the time and there was always a boiler of hot water on the open fire. Mrs Robertson said to Harry: 'Another hot day, Harry'. Harry replied: 'Cheer up lady, there's a cool change coming.' Mrs Robertson, in surprise, asked him how he knew. His answer was: 'Well lady, I was talking to one of those swaggies, and he told me that he had passed the cool change between Echuca and Deniliquin.' Harry Heath later retired and pottered around the station, doing odd jobs, such as sweeping up. His latter days were spent in the Deniliquin Hospital, where he died.

After shearing

After shearing, the wool was taken by the wagons into Deniliquin and loaded on the rail trucks to Melbourne, where it was sold. The Moonbria sheep were shorn at Boonoke even though they had their own shed but, having the big shed and every convenience at Boonoke, it was easier to shear them there. As some of the sheep had to come twenty to thirty miles, they were just moved a couple of paddocks every day so it was planned that the sheep were just brought in gradually and not knocked about. The same happened with the sheep at the back of Boonoke.

Grading firebreaks with old Austin tractor at Boonoke after 1931 flood

Shearing would cease by the end of August and in September all sheep were dipped in Cooper's Powder dip and the lambs weaned. The pasture in the paddocks for the weaners was kept short and well stocked during the winter, which is the growing period in the Riverina. This was done to keep the grass-seed in check; if the grass was kept short it reduced the incidence of seed. Corkscrew was the worst seed because it can kill sheep as it works through the sheep's skin. The corkscrew grass grew particularly well on the sand hills.

ABOVE: Horse team in wet conditions, Boonoke, 1931
BELOW: Wool on its way to Deniliquin

Rams

The ram lambs were drafted three ways. The biggest and most dense wool type lambs were separated from the weaner lambs that needed more attention and those with tails which were younger. They were then paddocked accordingly. Special bred rams had the best paddocks, same applied to the ewe weaners. Sheep were seen at daylight and dusk and kept out of any bad areas. They were moved on to water and troughs were cleaned out each day and at times twice a day. If they had to be wigged (wigging is the removal of the wool from the face so that sheep do not become wool blind) then we would muster them into the sheep yards out in the paddocks and do them with the

Boonoke North rams for show, 1930

shears. The boundary riders and some of the station hands could use shears well so if there was a big mob they would be taken out to help.

One day while talking in the sheep yards, wigging with blade shears, someone asked Zulu, a South African jackaroo, if there were many elephants in Africa. Zulu said that there were plenty and it was nothing to get up in the morning and go down to the toilet to find an elephant had pushed it over during the night, when rubbing its skin.

Sometimes there was a few of us wigging, so we would team up, one side of the drafting race pens against the other side, to see who could wig and clean the most sheep. However, this was often a source for argument. One day at lunch the conversation got to ways to count sheep. Some said twos and threes, others three and two, or as they come. One chap asked another who couldn't read or write how he learnt to count. He said that his old man would get a heap of sticks and throw them up one at a time. If I made a mistake he would whack me on the behind. Someone said that would be easy but the reply was that every now and then he would throw up a handful!

The young sheep were kept in those paddocks till about Christmas time. In January when the feed had reduced and the fire risk was less, the rams were moved over to the good Boonoke country, where the paddocks were a lot fresher. This was to push them ahead for March classing and ram selling. In January, all rams were wigged, pizzled (remove wool from around the penis), channelled (a small crutching), toe paring (the trimming of hoofs) and swabbed (swab with arsenic dip to prevent flystrike) with Cooper's dip in the paddock yards. Each ram was given two Cooper's pills so handling 150–200 was a good day's work. There was no drenching as the sheep didn't have any worms so it was only a precaution. Sheep were healthy as they weren't in paddocks that were overstocked. The most experienced sheep men got the stud sheep to look after. At classing, the sheep were always well grown and got favourable comment from the buyers. Classing was on 10 March and any rams not sold were walked to Boonoke North to be sold. It joined the rail siding so sheep could be trucked away. Ewes were classed later in June.

I had a great dog called Stumpy; it was a red dog and at classing time he was very good in the yards and always knew what to do without being asked, no matter who was with him. He had a lot of burrs in his coat at one time so I cut off the hairs and the burr. The cars in these days had foot or running boards and mud guards and Stumpy would sit between the two mud guards. On frosty mornings, he got frost all over his face and looked like a Polar bear.

Selecting rams for South Africa, 1927

Jetting

In the 1930s all the young rams and ewes were wigged and jetted in the open sheep yards. Lambing ewes were jetted in March for April/May lambing. The jetting plant comprised a thirty-gallon tank and a two-piston pump mounted on wheels, and an engine on wheels, which would drive the mixer by belt. A single jet – the filter, engine and mixing – was one man's job. It took two packets of dip, ten pounds, to forty gallons of water. This kept the ewes free from fly strike until lamb marking when the ewes would be channelled.

A ten-foot jetting race with galvanised iron sides was placed in an iron trough to catch the dip. Wood battens were put between the floor to make it a better floor and there was a little gate at the end of the race, to let the sheep out when jetted. A pipe was fitted to the trough to take away dip from the tray. This pipe took the dip to a five-gallon bucket which was emptied into the mixer. One man put sheep in the race as a decoy while one was being jetted and one was held ready to push in when the second sheep was finished, then the first let out. Two men kept the sheep up to the jetter, four to five men did the wigging and carting the water in the truck with a four-hundred-gallon tank. A mob would be about 1500–3000 according to which paddocks they came from. We would first throw a few buckets of water over the wigging pens to keep the dust down. Any sheep with seeds in their eyes were treated and a drop of lighting kerosene put in their eyes. This was used a lot for wound punctures, etc. The dressing was ten parts kerosene, one part spirits of tar and one part boiled oil!

Lamb marking

I stayed there for lamb marking in June (this is when all lambs' tails are docked, ears marked with registered mark and ram lambs castrated) and also helping muster the sheep into the yards. We would do 900–1400 per day as well as mustering the ewes. The team worked from the back-station using the truck if it was dry and the horses if it was wet.

With lamb marking, four men would catch the lambs and hold them on the rails while the overseer would ear-mark and cut off their tails.

We also used to crutch or channel out (clip away wool below the vulva to prevent wetting) all the ewes with shears at lamb marking time. Then they would be clean for shearing in July.

They would separate the wet from the dry ewes, which is taking the ewes that haven't had lambs out of the mob to give the ewes with lambs the best possible feed. The men would turn the ewes over with the overseer marking the wet ones with blue raddle and the dry ewes with red raddle. Then they would draft off the wet ewes. Wet ewes were channelled ready for shearing and any big toes cut.

All hands would clean up the dry ewes, two would be on feet cutting and three on channelling. Ewe classing was the next job and firstly ewes from Moonbria and Boonoke were done, then followed by ewe weaners from Wanganella Estate.

We started early and aimed to get all the sheep out of the yards by half-past three to give them time to mother up before dark. We stayed with the sheep till dark, going around them and seeing the lambs weren't getting in groups and splitting up from their mothers, and so the ewes could find them.

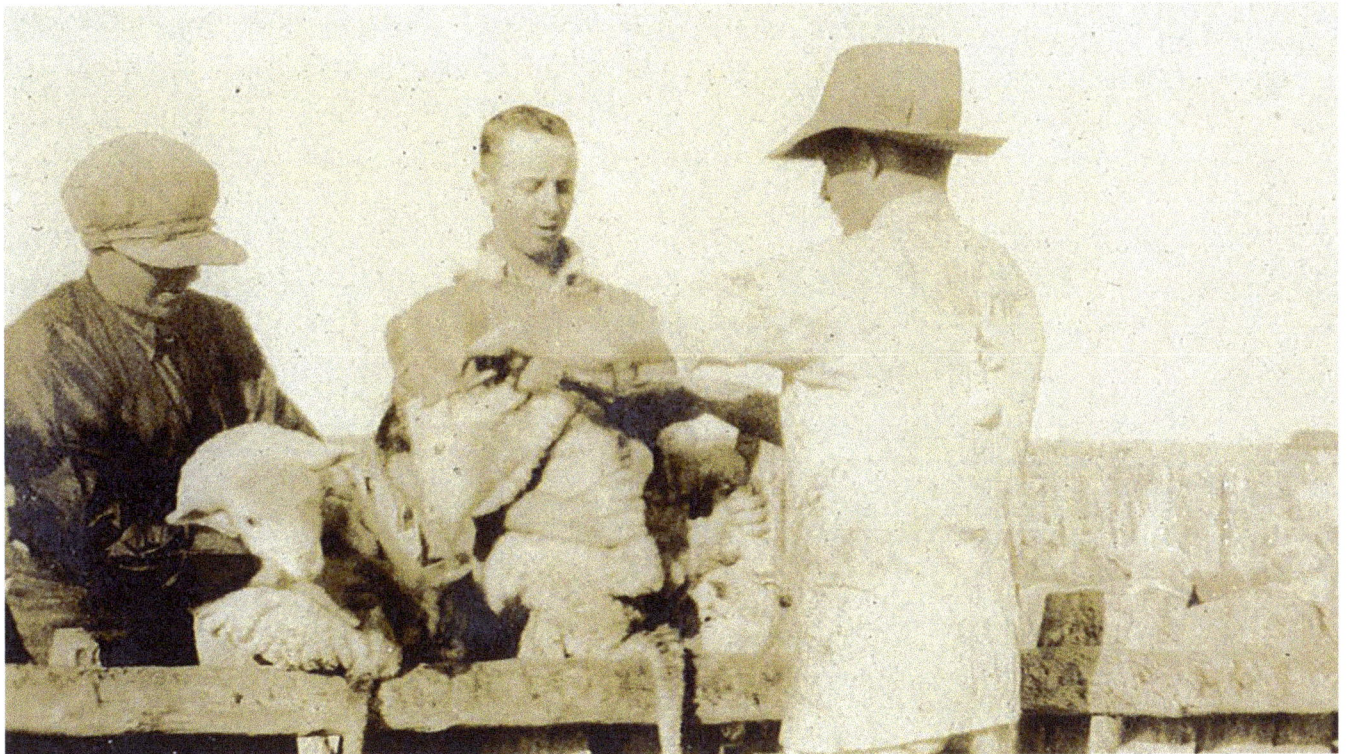

Boonoke lamb-marking team

The boundary riders (sheep men) wore out all the used shearing blades for paddock work during the year. They would wig four to five hundred per man per day in the yards, then a bit of channelling and maggoting (checking for fly strike). They all had dogs and carried dip and shears to do any blown sheep. When Bathurst burr was bad a burr hoe was also carried. On wet days one would oil or grease saddles and harness. When storms were about one would ride around dams to see if any drains were blocked and we would carry a shovel to clean them out if needed. One day I rode all day in teeming rain just doing this. They all had their lunch and tea bags in one side of their saddle bag, in the other, a quart pot to boil at a dam at lunch time. Only if the day was safe enough would you light a fire at the water's edge, some men carried cold tea.

Jackaroos were also known as boundary riders. It was said that the definition of a boundary rider was a station hand with no brains because after the lambs were marked and the dry ewes were cleaned the station hands went home. The boundary rider had to stay till dark though, to see the ewes and lambs mothered up.

I remember Mr Shallberg, the overseer, coming out and saying to Mr McMaster (who I called Mr Mac), 'I am taking Browser back in now that the lamb marking is over,' and old Mac replying to the overseer, 'Just when I get a good man you take him.' Mr Shallberg had been an officer in the First World War and spoke the same way to everyone.

Mulesing

About 1920 a new scourge reached the Riverina, one which was to cost graziers, and still does, millions of dollars every year: blowflies. Just as there were no rabbits to strip the pastures when squatters moved out on the saltbush plains, there was no such thing as blowfly strike. It, too, was imported. The species that attacks sheep is the metallic green Lucila cuprina and the Australian Encyclopedia says they came in from either South Africa or India late in the nineteenth century, slowly spreading inland to such places as the Riverina. There are still old hands at Deniliquin and Hay who can remember seeing the earliest struck sheep in the district.

The flies lay their eggs in moist wool, especially around the breech of ewes where the wool is dampened by urine. The maggots are fully fed within a few days and go to the ground to pupate. But the damage is done and struck sheep attract fresh waves of egg-laying flies. Graziers immediately had to start crutching programs to counter the blowflies, introducing a whole new factor. In time, one J. H. W. Mules developed a technique for surgically removing the folds of skin in the breech of ewes, and now thousands of sheep are saved by the Mule's operation.

In 1938, a Dr Kelly from Badgerys Creek came to Boonoke and he showed me how to mules sheep. He explained to me that the main point was to widen the bare skin on each side of the crutch, round the vulva, which keeps it dry. How they discovered this was, they had some sheep in the yards one day and there was one sheep that was plain, like a crossbred, standing beside a wrinkly sheep that had a wrinkly tail and loose skin down its hind legs. The wrinkly sheep was badly blown and the other, the plain sheep, was not struck with the fly at all. They caught those sheep and they noted that the wrinkly sheep had wool growing right up close to the vulva. The plain one had two inches of bare skin each side. Then they thought the idea would be to widen the skin each side of the vulva and it should give them good protection.

Dr Kelly also showed me how to vasectomise rams to use as teasers. A vasectomised teaser is a ram that has been subjected to a simple surgical operation to remove a section of a small tube called the vas deferens that carries the sperm from the testicle. This leaves the ram

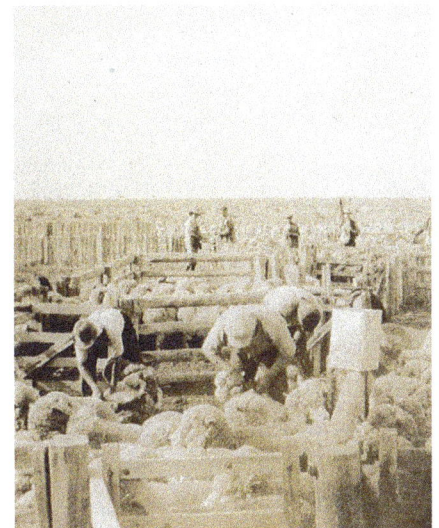

Wigging and toeing ewes at lamb marking, 1925

willing to mate but it is sterile. These rams stimulate the ewes to start cycling, then you put your good rams in with the ewes which makes a more even drop of lambs.

(Mulesing is currently still a hotly debated subject in Australia between animal welfare groups and those in the sheep and wool industries.)

Polo at Wanganella Estate

In 1925, Les Falkiner came to Wanganella Estate and Leigh Falkiner went to the French Riviera for a time. Les Falkiner was a good polo player so he and Neville Armytage, Ned White, Bob Landale and Douglas Boyd formed the Southern Riverina Polo (S.R.P.) Team and in 1927 won the Stradbroke Cup. The team that year was Bob Landale, Les Falkiner, Neville Armytage and C. Skene. They defeated Camperdown and Les Falkiner was the best player. They needed four more players to practise with and I was lucky to be selected as one; this was good, as I had most Saturday mornings off. We had to work most Saturday afternoons then as well.

Les Falkiner gave me a polo stick and ball and I would hit the ball a few miles out each day when riding round the paddocks. I would leave the ball and stick and hit it back when returning in the evening. This practice helped me play better so I was more help to the S.R.P. team. I would ride over to Wanganella Estate about five miles away where I changed into polo boots that Les Falkiner had given me. I had my own riding breeches, Indian cut too, from Pike Brothers in Brisbane. Indian cut breeches and jodhpurs were, as the kids now say, 'ace'. After a light lunch, I helped the Falkiners' polo groom lead the ponies over to Lindifferon, where Geoff Thornley had a polo ground in front of his house. We led two ponies each and rode one so there were six altogether. At Lindifferon, I was offered many ponies by Bob Landale and Neville Armytage. This was lucky because then I always had very good ponies. After the game, I would have afternoon tea and help lead the ponies back to Wanganella Estate and then stay for the evening meal. This was often followed by some entertainment as Mrs Falkiner was a good piano player and singer and Les a very good host. Then I would ride back to Boonoke, but often on the Sunday, I was invited back for tennis, so had a busy weekend.

Moonbria Cricket team, 1925

In the summertime, we used to have the odd cricket match, when we'd play either Moonbria or Conargo. If there was no fire risk we would travel into Conargo or Moonbria but otherwise we would play on the property in a paddock of Boonoke, near Conargo. The same happened when we played Moonbria; we would play at the boundary of Moonbria and Boonoke. We would have our fire equipment all there and telephone, too. There was a telephone line going along the road so we could just throw a wire over and hook on and get a connection. Boonoke North had a cricket team also and they played the Melbourne Cricket Club (M.C.C.) each year. Dr Kevin Cussen played for the M.C.C. His father, Sir Leo Cussen, was the longest serving President of the Melbourne Cricket Club. Kevin was to become my brother-in-law, as his wife Mary and my wife Winifred were sisters.

6 ZARA

Jack Bence, Zara; this horse was caught
at 9 a.m. and ridden by 3 p.m.

WHEN MY TWO YEARS AT BOONOKE WERE UP I went across to see Mr Thompson, the manager, to tell him I wanted to go back to Queensland. He asked me what I planned, and I said that I thought I might go out west there. He said that he thought I had had enough experience by now to take a position as an overseer and offered me one such, as the firm were thinking about buying Zara, another merino stud. I thanked him as I was very happy working for the Falkiners.

This is a local account of the sale of Zara, the pastoral property, to the Falkiners in 1927.

The purchase for £250,000, of the station, Zara, in the Wanganella-Deniliquin districts of NSW by Messrs. F. S. Falkiner and Sons, created widespread interest among pastoralists, particularly stud sheep breeders. This is the biggest transaction of its kind in recent years, and involves not only the land, but the stock as well. Falkiners do not intend to retain the flock, as they are of a different blood from those of the famous Boonoke line.

Zara has long been noted as a merino stud property, and its sale severs one more link in the chain of Riverina pioneering families, the Officers having owned the station since 1861, when Mr William Officer purchased it from the late Mr James Tyson. Falkiners' object in buying the property no doubt is to transfer portion of either the Boonoke or Wanganella Estate studs there, where the country is sweet and splendidly adapted for the running of stud merinos.

William Officer and S. H. and C. M. Officer, who were the original owners of Murray Downs, were the sons of Sir Robert Officer who, at one time, was assistant Colonial Surgeon in Tasmania. He was later a member of the Tasmanian Legislature and became speaker of the Legislative Assembly. William Officer started a Tasmanian blood stud in 1889, which he continued at Zara until 1913 when he changed over to Peppin blood stud because of their better performance. His son Ernest took over running Zara in 1913 and continued until 1927 when he sold to F. S. Falkiner & Sons. He had been the first president of the Windouran Shire Council in the Riverina region of New South

Wales. Ernest retired to Toorak where he died in 1936. (Ironically my daughter would end up marrying a relation of Ernest's, Donald Officer.)

Ernest Officer had requested in his will that his ashes be taken by plane and distributed over Zara. This was duly carried out. In the 40s, after the 1943–44 drought, Mr Otway Falkiner said to me, 'Browser, you know, since we have scattered dear old Ernest Officer's ashes over Zara, we have had nothing but bloody drought!' We'd indeed had some lean years then.

Mr Shallberg went to Wanganella Estate in 1926 as manager, and I went to Zara as overseer after the Boonoke shearing. This was a job usually for one in his thirties, so I was fortunate as I was only twenty-two years of age. F. S. Falkiner purchased Zara for three pounds three shillings per acre, as well as a good stud of 1600 sheep, which were sold on 16 February 1928. These original Zara stud sheep were sold by the firm of Wilkinson and Lavender, Hay. After the sale, a charcoal drawing of the sale was left in the boss of the board's cottage. It was very good and I still regret not collecting it. It was done on thick paper and hung in the dining room at the homestead, but later it fell and was damaged.

Boonoke horse teams, 1924, loading wool to transport to Deniliquin

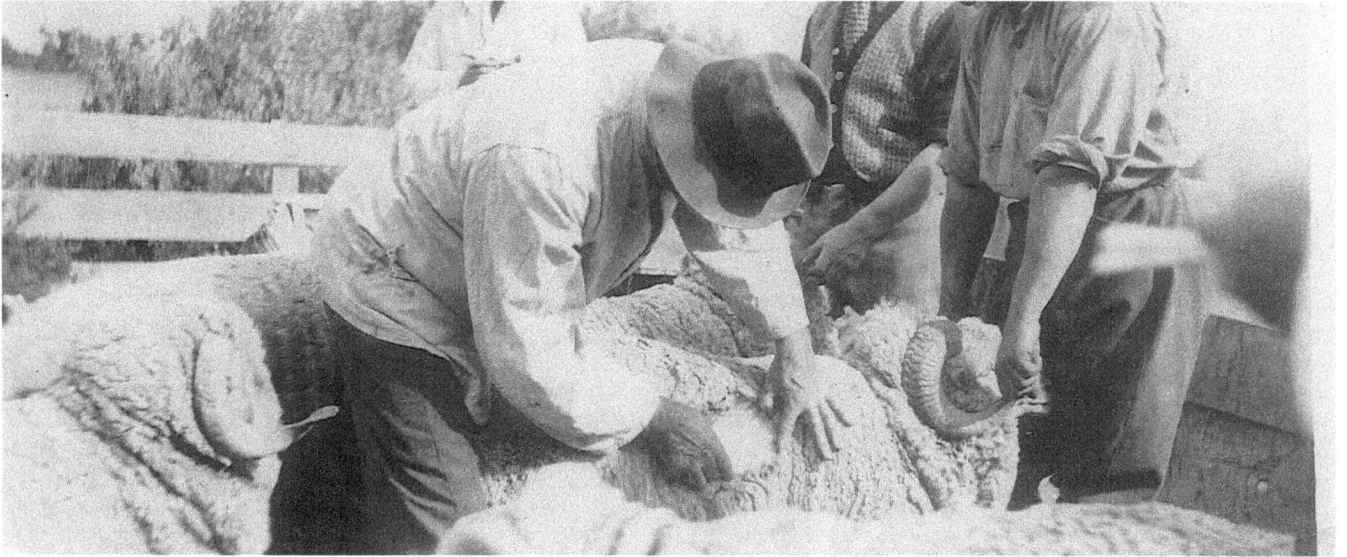

Pennyfather Snr selecting Zara rams
for South Africa, 1927

That night after the sale there was a good party at the Zara homestead with the agents, some buyers and Falkiners, as well as some staff there. The old chief (Mr Otway) was mixing his Billabong Snifters which were powerful drinks; just one would give most people a kick. He congratulated his son Les on the way the sheep looked, and on the sale. Les said, 'You don't want to congratulate me, Father, you should congratulate Browser.' So, when Mr Otway came over and told me this I smiled. He then said, 'Don't laugh at me, I mean what I bloody well say.' This always amused me because instead of just getting praise I got a chip as well. I knew him so well that I was not offended. About 1929, Les Jowett, who was one of my jackaroos when I was overseer at Zara in 1927, was made overseer at Moonbria.

Sheepdog trials

Les Jowett was a good hand with sheep dogs and he ran a sheep dog trial at Moonbria. He got John Armstrong and George Guy, two old experienced men, to be judges. It was held near the woolshed and proved to be an interesting day. The overseer from Steam Plains won the trial with a black and tan dog called Smoker (afterwards I bought the dog for ten pounds, a big sum in those days). The trial was discontinued because Les Jowett left Moonbria but then a trial was started in Deniliquin at the annual show. My dog, Smoker, was working in it until he heard shots at a shooting gallery, and took fright and got under a car.

Jack Bence and two Zara jackaroos

7 FLOODS, FIRES and OTHER TALES

Men going to repair shearing machines in the 1931 flood; rouseabout huts in the background, Boonoke

Floods

The year 1931 was the year of the big flood, when there was water from the homestead to the Forest Creek, on the Deniliquin/Conargo road. Mail or passengers for Boonoke had to be brought in by horse and gig because the water was a few feet deep. When the Auditor from Melbourne came, Mr Thompson told me to send Plugger Bill, the teamster, to pick up the Auditor, Mr Jackson, and take a coil of rope in the gig. Mr Jackson was a real city person and of course asked what the rope was for. Bill replied, 'The boss told me to take the rope in case I had to put it round you when we go through some deep water.' Consequently, Mr Jackson naturally was very scared.

One day three or four of us were riding out to do a job and had to cross Browns Creek at the yards, where there was a good crossing and the horse only had to swim ten feet. A jackaroo, Tom Lyons, had on a pair of gum boots and when crossing he had one hand on his saddle bag to keep his lunch dry, and his feet up so his gum boots would not get full of water. Unfortunately this tickled his horse's flanks so that it reared up and pig-rooted. Tom went down head-first, with his gum boots still above water. It was funny to see two gum boots floating above the water. Unfortunately, Tom was killed in a car accident in 1943.

Returning with the mail, 1931 flood

Once we were building a bridge across a deep gully near the woolshed, which was full of water, and Evan Cameron fell in. He could not swim so crawled out with his eyes shut. He kept crawling for about ten feet before he realised he was out. We all laughed at his misfortune.

The flood was just starting to spread out from the Billabong when two station hands, Jack Edwards and Jack Pitman, drove a gig to Deniliquin to get clothes as they knew they would be isolated for some months. They stayed a night in Deniliquin and returned the next day. On the way they called at Conargo, where one chap lived, intending to arrive at Boonoke in time for 6 p.m. tea. They found when they got to the Billabong that it had risen quickly and spread over the road in varying depths. The track was heavy and after a mile or so they thought that the gig was too heavy for the horses. Therefore, Jack Edwards led the horse and Jack Pitman got out of the gig. After a bit of pushing he took off his boots to make it easier walking in the mud. That night I was in the office with Mr Thompson at about 9 p.m. when a chap came into the office and said someone was coming up the creek. So, I jumped on a buggy horse at the

TOP: *Boonoke woolshed, 1931*
ABOVE: *Evan Cameron checking the channel break*

91

Partly loaded wagon bogged on Boonoke, 1931

stables and went to help, taking a length of rope with me. I kicked my shoes off at the water's edge and found the two making slow progress about half-way to the bridge. The water was about three and a half feet deep but they were OK so I just rode back with them. The next day Jack Pitman had to go to bed with swollen feet from prickles from the creek bed.

That year shearers and shed hands who did not have horse transport were brought in on a wagon and wool was carted out from the shed to the road in wagons in loads of twenty bales. They were then loaded onto other wagons to go to Deniliquin. I think about six hundred bales were in the shed at the finish of shearing. Due to the flood the wild oats were tremendous and so were the marsh mallows. Cattle were brought in to knock down grass and make safe areas. The firm bought a lot of weaner cattle from Newcastle and northern New South Wales, which were then fattened and later sold. Some paddocks at the back of Boonoke were not stocked until after Christmas in very good years until the feed reduced a bit. Falkiners never overstocked.

Fires

In a fire risk period sheep were put on water soon after daylight. After three or four days they were trained to go to a water camp until it was cool and time to feed. We would then just check each day that there were no fly-blown ones in the mob and that the troughs were all right. If we did have to catch any sheep, we would round them up near the dam or trough so they would trample the grass down and make a safe area. After a few weeks, there was quite a big area trampled or eaten out, and safe even if a fire did come.

As Boonoke had such large plain paddocks it was the biggest risk area. The men were up early and would be back at the outstation or the homestead by one o'clock. No smoking was allowed when riding paddocks and anyone who wanted to smoke was to stop and have it at a bore, creek or dam. They would then have the afternoon off because they had been up early.

Grading firebreaks on Boonoke, 1931

Grading firebreaks with an Austin tractor on Boonoke

Because of the mustering, the manager and all the men knew where all the sheep were. In the office at Boonoke, there is a big board with a map of Boonoke showing where all the fences, dams and mills are. We would put onto the board a pin with a flag showing the number of sheep in each paddock; the bookkeeper also knew where all the sheep were. If anyone came and there happened to be a fire, the bookkeeper could direct them from the homestead to the fire.

There was also fire equipment on hand. Two-wheel Furphy fire carts were pulled by a horse. The cart carried a hundred-gallon tank of water and had a hand pump attached, plus axes and fire beaters. The manager had a little Ford ute with a fifty-gallon water tank and the overseer an old Napier cut down to ferry water if needed. The only other station equipment were wagons, drays, sulkies and carts. Some bag fire torches were used for back-burning by our wetting them with kerosene. By the thirties, Falkiners had a six-ton Thornycroft truck and a two-ton Dodge truck, and both the manager and overseer had Dodge utilities. Trucks were kept in readiness and all fire equipment painted red and left on trucks for the fire season. The equipment was bigger and included an Ajax water pump but all work was still carried out by horses and bullocks pulling wagons, drays, carts and buggies.

Most fires started in the afternoon. F.S.F. supposedly lost only twenty-six ewes in the big fire of 1918. Thousands of sheep weren't so lucky on other places. It apparently burnt from the Murrumbidgee to the Murray, jumping the Billabong and the Edward. It went through the Puckawidgee part of Boonoke and Mr Thompson who was there told me on the western side he put out the fire himself. He only had a little Tin Lizzie and a fifty-gallon fire tank, a few fire beaters and one man with him. The wind was so strong and it was coming from the north when suddenly it changed to the west and blew out the western side of the fire. They were lucky but as well all the sheep were on water. Mr Falkiner had heard the news in Melbourne about a terrific fire in the Riverina that burnt thousands and thousands of sheep. He came up to Deniliquin straight away on the train and when he was met by Mr Thompson, he asked, 'How many thousand sheep did you lose (he had heard that nearly all the Boonoke flock had been lost)?' Mr Thompson answered, 'Twenty-six sheep, old ewes.' Mr Falkiner's reply was, 'I'll be buggered. You must have lost more than that.' It was only that these old sheep were in a far paddock on Dormie and hadn't made it to the water and that is the reason they died. Fortunately, in my time there were no big fires. There was one at Warriston but we got it out without much damage.

Other tales

Kangaroos were controlled by having a shoot every three years. Anyone with a shotgun and a ute was invited, cartridges were supplied plus a paddock lunch with beer. We reckoned we would shoot five hundred each day and there still appeared to be a lot left after the shoot. We went through Moonbria one day and Boonoke the next. Shooters were followed by a truck and some men who picked up carcasses and put them in a heap to burn. This stopped the carcasses rotting and breeding flies.

Old Jack Wiley was the rabbiter at Boonoke. He had a large pack of dogs and rabbited every day of the year, working over the entire area. He asked Mr Thompson once if he could do something for him. His reply, 'Yes, Jack, as soon as we get some spare time.' Jack Wiley said, 'Mr Thompson, I've been with the firm about twenty years and I've not seen any spare time yet.'

One day the overseer asked me if I could kill a beast. I had helped in Queensland, so, full of confidence I said yes. So one afternoon, Jimmy my offsider and I set to work on a bullock which had lumpy jaw caused by an infection in the mouth started by sharp grass seeds. I thought one shot in the curl on the forehead would be enough but it took three shots to drop the bullock. A chap at the homestead heard the shots and said he thought it was a shooting gallery. Afterwards I found out quickly that the right spot was two inches above a central line between the eyes. Another mistake I made was when I failed to cut the pizzle off a bull and the stupid butcher/milkman rolled it in pastry. It went to the hut and when one of the men was carving his helping at dinner that night he yelled to the other men, 'I've got the bull's pizzle so hold on, we may yet get its balls'!

My first visit to Deniliquin was when I needed to go to the dentist. I think it was the summer of 1925 and riding then in the 'twenties was normal, so I rode my horse in. The dentist, whose rooms were opposite where Brian McCleary & Co., the local accountant, is now End Street Deniliquin, was an old bloke, who had the old foot pedal drill for drilling, just as good as the electric ones, but slower. In those days, the shops had hook-up rails on the street to tie up one's horse. I remember stopping at the Sportsman's Arms Hotel for a beer before leaving town and giving my horse a drink at the large horse trough. It was over one hundred degrees so we both needed some refreshment for the journey home.

OPPOSITE: *Men swimming Browns Creek, mustering for Boonoke shearing, 1931*
TOP: *Plugger, Jack Edwards and Pat carting dead wool*
LEFT: *Gat with his dog Nigger, Nindigully,1923*

8 HORSES

Gat marking a colt in Boonoke yards, 1926

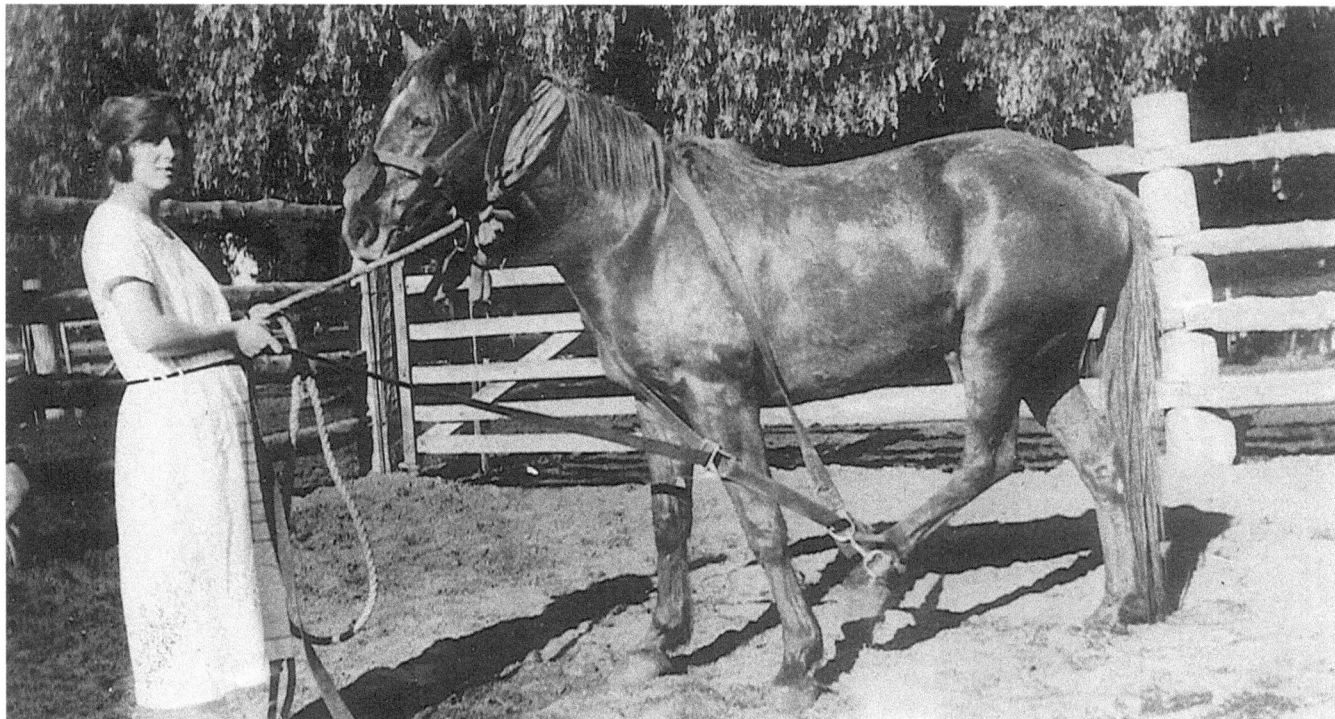

Otway Falkiner's daughter Emily Falkiner holding an Arab pony being broken in at Boonoke North

F. S. FALKNER ALWAYS HAD GOOD HORSES. Boonoke used to breed foals every year as well as hacks and cart horses. The horse-breaker would then come when the foal was old enough to wean. He would handle the foals and break in the ones from the year before, breaking them in as two-year-olds. Boonoke used to run twenty to twenty-five well-bred mares and a thoroughbred stallion. At weaning, branding and marking time the breaker had done it all himself until I came to Boonoke.

Up in Queensland they used to talk about marking colts, so I was quite keen to learn how to do that. One station hand had a piebald colt, a two-year-old, and I said to him, 'If you let me mark your colt Jack, I'll give you five pounds if I kill him (more than he was worth).' Mr Thompson heard about this and said, 'You would like to learn how to do colts, would you? Well you can do them all.' From then on, I did the lot, and I was the Boonoke vet as well for the surrounding area. At the time, there was only one vet in Deniliquin, and once when he was sick he asked if I would go and mark his stallion for him. I was quite pleased, as it was a very well-bred stallion, Colonel Worthington. Mac Falkiner, Andy Hermiston and I went over there and threw him and marked him. I thought it was a compliment being asked by the vet.

Luckily the breaker and I were friends and he told me all he knew, so much so that he wanted me to go away with him breaking in horses. He offered me a job at three pounds a week clear, which was very tempting. I did a lot of thinking and decided to stay with F.S.F. as I knew the main things about handling unbroken horses. I thought it was better to get more general experience, which proved a wise decision, as the following year I was made overseer at Zara.

The horse-breaker, Jack Bence, was a bow-legged champion rider, who was an excellent horseman as well as bushman and vet. He did a lot of the main stations in Victoria and the Riverina and had his own way of handling horses. Kinnear's in Melbourne made special gear for him. The Bence gear made handling foals and older ones much easier and didn't knock the animal about. Jack showed me the way

Jack Bence

to mark colts was for one to hold the colt by a halter and the other to put a rope over the neck and round the rear hind leg above the hock. It needed to be a slip knot that tightened when the one standing near the colt's neck pulled the rope and over it went. When it fell, a bag of sand was put on its neck. This was better than the old way with two ropes and four to six men: having only one leg tied, it is paralysed by the knot.

When Jack returned from breaking, he did only a year as he was engaged by the thoroughbred breeders of racehorses. He was to work mainly at Caulfield to handle young racehorses. I did meet him once after that at Flemington in the trainers stand. He told me that he would give me all his gear. I thanked him and told him how I appreciated his gesture but really did not have enough use for it.

Emily Falkiner with a pony at Boonoke North

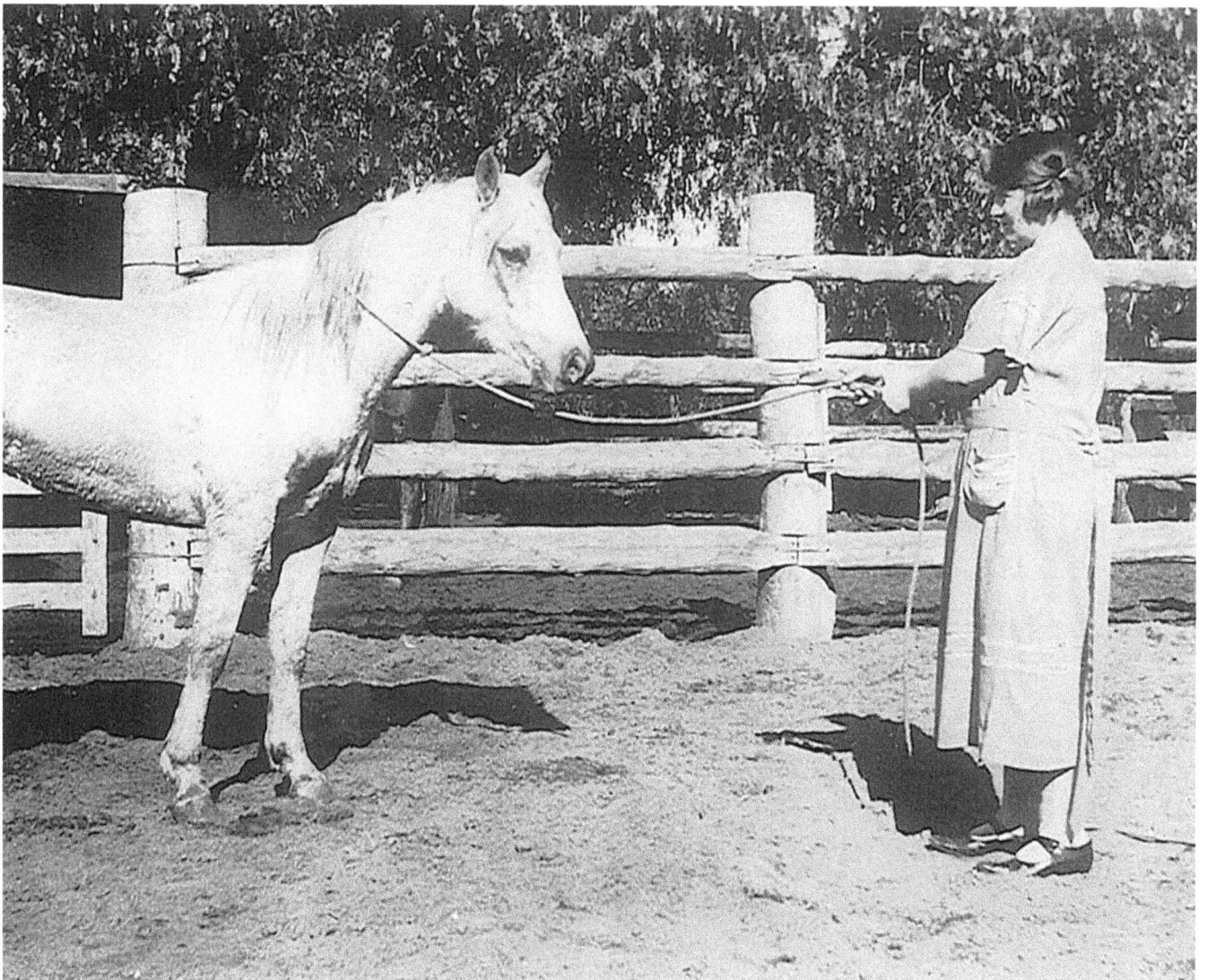

Vic Cowan took over breaking with F.S.F. when Jack Bence left and he too was a fine character and horseman and represented Australia in America. I remember a jackaroo/bookkeeper boy about twenty years of age was once about to mount a horse and I said, 'You can ride, can't you?' and he said, 'Of course I can.' The next thing I knew he mounted by putting his toe into the horse's ribs and was quickly put over the horse's neck.

Mr Otway Falkiner's daughter, Emily, would stay at Boonoke and liked to have her own pony to ride. Emily was feminine but she could have taken charge of sheep. I was asked to ride her pony before her visit. I rode from Boonoke North to Boonoke, about eighty miles, but fortunately with all the directions I made it in a straight line. The first day I got to The Yanko where I stayed the night in The Yanko barracks (getting all The Yanko gossip). That night for the meal they were having ducks, which the jackaroos had shot. The bookkeeper had a crow, however, as he was always saying how good they were to eat, so they thought they would try him at his word. He ate it too, without comment. The next day I got to Moonbria for lunch and back at Boonoke for afternoon smoko.

At Boonoke North, they had a racehorse stud. David was bred at Boonoke North and won the Sydney Cup. Afterwards he was used as a sire. He was a good one and produced many winners in Melbourne and Sydney, particularly in jumping races such as the Grand Nationals. They all had a good trait, being quiet and intelligent and easy to work. Mr Otway Falkiner gave Jeanette Falkiner, Mac Falkiner's wife, a good two-year-old colt by David. He was a beautiful horse; all of David's offspring were quiet-natured horses, too. She called the colt Saul and it won a Grand National in Melbourne. There was another one Jeanette owned called Zalmon. It also won a Grand National Hurdle.

At Boonoke North they also had a big Clydesdale stud at a part they called The Farm. Two notable sires were imported, Baron Belmont and Craigie Master. I think Craigie Master came from Scotland along with the overseer, Willie Mathewson, who was a small man. He came to only half-way up the horse's shoulder, so he looked funny controlling such a big horse. I happened to be the vet for those and I used to do it the bush way. I'd mark them in the paddock, then let them go, rather than doing them in the yard where there was risk of infection. I had them taken down to a paddock, three or four miles from the homestead, along the Billabong and the sand-ridge. Jimmy Waters, who used to be a trainer, was there and led them down and he and I popped them over on the sand hill and marked them.

A bit later, I had an idea about bobbing tails. This was done by cutting a joint or two off the tail end. When I tried this two joints seemed so little, I hardly thought they were worth doing, so I took another joint off. That made it look a bit bob-tailed, so much so that it was noticeable. The idea was to make the horse look square behind instead of being narrow. The horse in question was called Browser, as a bit of a joke. Browser went to Melbourne and had won a few races down there. One day it was running in a hurdle race and fell. The next thing we heard was that Browser had to be shot! It was on the radio and my wife, Winkie (see introduction), was listening to the race which caused some angst.

Jockey Jack

In the 1930s we had a groom known as Jockey Jack, who came to Boonoke after working as a stable-hand and exercise jockey at the Deniliquin stables. He did his job well, looking after the stables and the stallion in the loose box and running the mail twice a week to Conargo. He was only jockey size but could defend himself. I remember one morning after orders he was speaking up to the horse teamster, Plugger Bill, who was a tough, rough chap, and talked of guzzling people. Someone said, 'What would you have done if Bill had turned on you?' He said, 'Well, I could always fall back on my fork!' (This was the short-handled pitchfork he used to turn over the straw in the loose boxes.)

I think I read in a magazine called Hoofs and Horns about Jockey Jack and 'he' was apparently a girl, who had come out from England about 1920. She could not get a job as a girl in Sydney so she changed and acted as a male and then was employed. I met 'him' again at the Wanganella Hotel when the Hay coach arrived. I was picking up the Zara mail during the Second World War and he was in uniform, taking some internees to the POW camp in Hay.

9 FROM 1930 to 1939: MARRIAGE

It was not until 1930, when Mr Thompson lured me back as assistant manager, that I returned to Boonoke and lived at the homestead. I had a room next to Mr Thompson, which was joined by a door between our rooms. He had a bad heart. He wanted me to be there so that if anything happened to him, I could be left with the old chief to carry on. We became very close friends. Mrs Thompson was at the time in hospital in Bendigo. Often late at night Mr Thompson would call me into his room and we would talk about his past experiences. I gained a lot of knowledge from him.

In my opinion there was no better station manager than Mr Thompson. He was a wonderful character, a gentleman with a wealth of experience, and one of the finest station managers and characters I had ever met. He was passionate about cricket and encouraged Boonoke to have a team and to play the other stations.

Gat, Derr Forster and Gat's sister, Louey, going to a test match at the Gabba in 1930 to see Don Bradman bat

He was an excellent stockman and some even said Mr Thompson was half the reason for Boonoke's genetic progress. Otway Falkiner usually had a long discussion with him about the rams. He often carried on with classing or mating ewes while Otway was looking after clients. Thompson's forte was an incredible touch and judgment when handling huge numbers of sheep. On Boonoke, it was especially critical with mobs like weaners and knowing when to move them and what paddocks or properties to use well ahead of time. This enabled Boonoke sheep to be always well grown, hence the one-year-old rams at classing were always very big. I remember the first morning after returning to Boonoke, I had to give orders to the men, as I was now the overseer. These were men I had been working with before as 'Browser' but when you became overseer you had to be addressed as 'Mr' because in those days everyone in a position had to be called 'Mr'. So I said to the men, 'From now on I would like you all to call me "Mr Gatacre" as I am your boss, but you can call me whatever you like off the place." '

I had a 7-shot Colt automatic revolver, which was given to me when the secret army was formed. The New Guard was formed in Sydney, Australia, in February 1931 as a paramilitary offshoot from a conservative tradition defending loyalty to King and Empire, sound government, law and order, individual liberty and property rights. In particular the movement was formed in response to the policies adopted by Jack Lang, the leader of the Labor Party and Premier of New South Wales. It was led by Lt. Colonel Eric Campbell, a First World War veteran. The New Guard declined rapidly following Lang's dismissal in May 1932 with its remaining members becoming increasingly inclined towards Fascism. Still led by Campbell the movement unsuccessfully attempted to enter parliament at the 1935 state election (running as the Centre Party) but disbanded completely shortly after.

The secret army was for the protection of the station property. The manager at Boonoke and I were the only members and our superior was Mr Armytage at Quiamong. Mr Thompson, the Boonoke manager, and I went to Willurah and enlisted with Mr Lamb (Harold) and his overseer. This was as much as I knew and that was how the whole country was organised. Fortunately, conditions were peaceful. I

Louey and Phyl Pole, Brisbane races

practised at odd times with the revolver and my best shots were when going fast in my ute towards an eagle on a post and shooting it as it took off, and shooting a snake swimming across the creek. I gave the revolver to the police when we moved into town from Wanganella.

My brother and his wife stayed with us, when he had some leave during the last war. We all went shooting along Browns Creek for ducks. My brother's wife, Wendy, aimed at some ducks flying over and shot a black duck through the head. I would no longer talk about my shots! Don Officer senior got two ducks and thirteen painted snipe with two shots. 'Jum' Falkiner once shot a swan overhead thinking it was a black duck, though it was after a celebration lunch!

Mac Falkiner and I had a lot of good times together during our seven years in the Boonoke house. It was before neither Mac nor I were married so we were like Mum and Dad. Few people knew that Mac had a blind eye which he got in a cricket match at Boonoke North. It was not noticeable and I can tell you he saw more than most people with two good eyes.

Mac got married in 1934 to Jeanette Cullen and they had a big wedding in Melbourne. It was a day wedding and we all had to wear morning suits and top hats, which was all new to us. We had to hire them, of course, so we went around to The Australian Club where a lot of the members had their hats hanging up so we tried on different ones to find ones that fitted us. I remember I had Sir Hugh Gipp's hat. When we returned to Scots Hotel we thought we were looking good but the lady at the desk said to me, 'Monsieur, your hat is far too small for you.' Mac and Jeanette insisted I stay on in the house when they returned from their honeymoon, so I was part of the family.

This arrangement continued till Mac and Jeanette had a family. They had quite a few mothercraft nurses to look after their children, Angus, Sally and twins David and Ian. In March 1939, Winifred Stokes arrived from Echuca as the new mothercraft nurse. A friendship soon developed and I fell for her and we got married in October 1940. After we were married we lived in a house overlooking the Billabong Creek, just along from the Boonoke Homestead.

Winkie's family, 'The Stokes', at their home in Echuca

10 FAMILY, DROUGHT and WAR: the 1940s

*Philip was born in the Stokes' home
in Echuca on 26 July 1941;
Gran Stokes, Philip and Winkie,
with nurse, July 1941*

Family and war

IN MAY 1941, I WAS MOVED TO ZARA as manager and my wife and I lived in the Zara homestead. Half the homestead was the original, which was built by the Officers and the other half was built by the Falkiners from bricks from the Wanganella Estate homestead, after it was burnt down. This was of course war-time and we were going into a drought. Although we didn't move much from Zara, we had quite a few visitors. During the war years, we had the house full most of the time so we had some help in the house consisting of a cook and two housemaids. The two housemaids, Ester Williams and Anne Shields, decided one weekend to visit the Cooks Ridge outstation in the middle of the drought. A severe dust storm blew up, reducing visibility to zero; they got lost and were not found for two days. We had flower- and vegetable-gardeners at the time as well.

Gat and Winkie having lunch at the Deniliquin Picnic Races, 1946

Gat, Winkie and her sister, Betty Elliott, at Deniliquin Picnic Races

Mrs Leigh Falkiner and Mrs Ralph Falkiner and their friends came to stay and Mr Leigh Falkiner left us a nice present in his will of one hundred pounds in appreciation of the way we looked after his family. My brother Galfrey's wife and family from Sydney also came when the Japanese subs got into Sydney Harbour. People in Sydney who could leave were advised to go to the country areas. My brother came to spend his leave with us, before going to the Solomon Islands naval battle. An English Admiral and his wife, Mrs Pizzey and some other naval officers spent time with us, including one officer on sick leave. Our family helped quite a few during war-time by providing hospitality, friendliness and relaxation. My wife played a big part in this. Our main source of socialising were the tennis parties shared between ourselves, the Austins at Wanganella and the Macaws of Caroonboon.

One social event we always enjoyed was the Southern Riverina Picnic Races which were held annually; the first day's racing was followed by a Dinner Dance, then the second day's racing followed by dinner and a formal ball. We would have guests to stay in the house. The races were suspended for both the First and Second World Wars.

There were numerous stories associated with the Picnics and one I remember well was of a friend of ours going home and pulling off to the side of the road to have a rest. The car got caught on a slight

Philip on Timmy, Zara, 1946

mound with one wheel not touching the ground, so when he went to take off, he found he was not moving. So he fell asleep again. Someone pulled up and went over to see if they could help. Tapping on the window they asked the driver how he was going; he woke up, looked at the speedo and said he was doing about twenty-five miles per hour.

Our first son Philip was born in 1941. Our second child, Louise, was born in 1945 and five years later, Mike arrived. We were at Zara for ten years where we were raising our family. We didn't get about much though, because first we had the drought in 1943–44, the war was on, too, and there were shortages and petrol rationing. We had a little Morris Minor, an eight-horse-power ute, which did the job for us. It did shed a tyre at times though, such as the day I was taking my wife to Deniliquin to have our second child, Louise. But everything went well and fortunately, Win isn't an excitable person; I did not fancy trying to be a midwife on the Hay road. The speed on the roads then was thirty to forty miles per hour, which was fast enough on the old dirt roads. Our only other social visits were going over to Wanganella to see the Austins or the Austin children would come over to see our children. During the war Deniliquin was an important

Gat, with Louise on Spider

centre for the Air Force and had a big training base. One day we had a forced landing in a paddock close to the Zara homestead; fortunately no-one was hurt. They had a night with us at Zara, got the plane fixed up the next day and flew it back.

A lot of places had charcoal-burner cars but we managed with the little petrol car which we only used to go to Deniliquin about once a month. We used horses for all station work though. We were lent a Shetland pony from Caroonboon for our son, Philip, to learn to ride. The pony was certainly cunning though: as if he had had enough of Philip on his back, he would often head under an orange tree and lose Philip in the process. Like a lot of Shetland ponies, Timmy foundered. To give himself relief he would stand in the creek for hours, up to his belly in water. After seeing him do this we kept him yarded at the stables and fed him lightly for a week or so and this was the cure. Philip was then aged six years and it was not long before he

Gat on Spider, Philip on Timmy, Zara, 1946

could ride a pony; they aren't as cunning as a Shetland. Timmy was returned to Caroonboon where he spent his final years in retirement and where he was also buried, when he eventually died, aged thirty.

When John and Jill Dickson decided to sell Caroonboon in 2014 after the Dickson family had owned it since 1861, John was asked about his time at Caroonboon. He said he had vivid memories of his childhood, especially his Shetland pony, Timmy. 'Timmy was a mischievous devil and we were always in trouble,' John said. 'In those days, we were reasonably isolated and there was bulk flour and tea and sugar in a store at the station. I would open the lid of sugar bins to let Timmy have some and we would both get into trouble.' Timmy also taught a host of children to ride at Caroonboon.

Drought

The 1943–44 drought was the worst drought that I can remember and older people than I at the time couldn't think of a worse one either. It looked in 1943 as though that was the drought year and we all expected it to break in 1944. Instead that was also a drought year, so we had two drought years running: three inches of rain in '43 and four inches the next year. It didn't break till 1945 so it was hard keeping the stock alive. We kept the feed up to them in 1944 by getting grain from everywhere and train-loads of vegetables came to Jerilderie to supplement their feed. This kept a lot of the residents of Jerilderie fed for a while too.

Zara horse-yards, cart, gig, shed and stables were old but in good order. At the yards, there was a big underground iron tank with a top about fifteen feet wide. This was closed in by a rail fence and the water was used for the stables. A bit of grass was growing between the fence and the tank and in the two-year drought grass was scarce. Therefore, the gig-cart horse, Sox, and the night horse were fed and kept in the yards. Well, one evening old Sox pushed his way into the enclosure for the few blades of grass and fell on the top of the tank. He pushed it in and poor old Sox was in the water. The noise was heard by a contractor camped by the stables and we all came over and began a rescue mission. Old Tom, the blacksmith, who used Sox to get the mail from Wanganella twice a week, got a rope and put it around Sox's neck. He had to hold it in case Sox's head went under the water. Then all hands went to work – a great experience for them.

In 1944 we sent a lot of the sheep away, up to Coonamble and Rowena by train. I would often go up with the sheep on the train to northern New South Wales. I would stay until they were settled and then come back. Before trucking at Jerilderie the sheep were fed on vegetables while yarded overnight, as hay was not available anywhere. It was a seventy-two-hour train trip and we didn't unload them for any stops on the way. It was thought unloading them and giving them a feed at a station, then reloading them, would knock them about more. When unloaded, they were given Mitchell grass and hay, before the drover started to get them on to green grass and herbage. We only lost a few sheep which was good. I also made other trips to check and attend to the sheep. F.S.F. leased a place called Pimpampa, which belonged to the Jaquets and the only sheep there were the Falkiner stud ewes. I stayed with the Jaquets who became good friends to me. When the sheep returned, they were fat but covered with a clover burr, which was bigger than the trefoil burr. While in the trucks they ate it off one another so when they arrived at Jerilderie they looked lousy, as they had bits of wool hanging where they had chewed the burr off.

George Falkiner of Haddon Rig started making a film about the story of a sheep's life, called 'A Lamb for Mary'(1948). This began with the lamb's birth, then weaning, grazing, enduring drought and good seasons after rain, shearing, manufacturing and ending as beautiful gowns on mannequins. On one of the train-loads going up to Pimpampa, we had the film crew with us. They included Alec Cann, who was an American Army war correspondent, and was the producer. There was also Doc Sternberg who was the editor (I think he later became manager of Her Majesty's Theatre in Melbourne) and Karl Kaiser who was the photographer.

This was an experience, being with these fellows while they were filming. They used to discuss, on the train, what they were and weren't going to do in the film. One time I thought they were having a big fight but they were just discussing the film, even though they were calling each other all the names under the sun. I appeared in the film in one part, giving delivery of the sheep, returning home after the drought. I was counting them out from a paddock to the drover and as the flies were bad I kept swatting them. My sister saw the film a year later and she said, 'I thought you were just swatting flies, not counting sheep.' When we were at Coonamble inspecting ewes, we found three Border Leicester rams in a mob. Alec had his army carbine rifle with him so we shot and then burnt the rams so there was no evidence! These were the stud ewes and we did not want them joined and impregnated by unsuitable rams from the neighbours.

After this trip, Cann and his two offsiders, Sternberg and Kaiser, came up with me on the train. When making the film, other areas were filmed at Haddon Rig and Boonoke. I had the use of an old Chevrolet ute at Quirindi when staying with the Bensons. I would go to Quirindi by train and then on to Coonamble, Rowena and Burren Junction by ute. Bennie

Vic Brenanan and Robbie about to go for a Sunday drive in the windmill ute

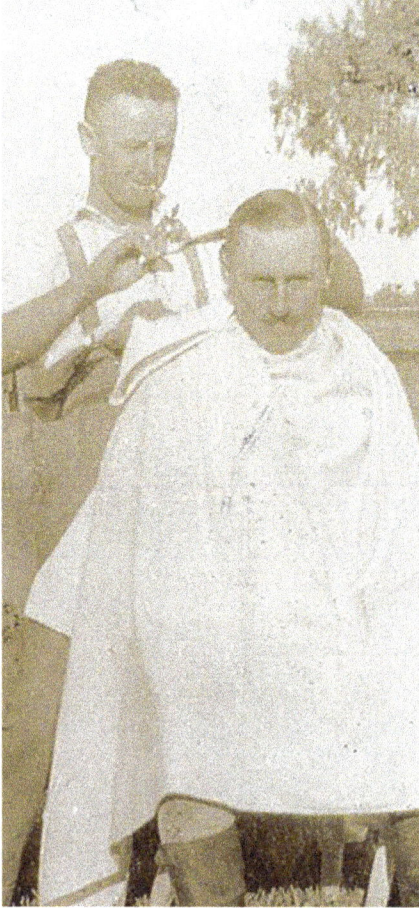

Vic Brennan giving Robbie a haircut

Benson used to manage properties for NZL (New Zealand Loan?) and knew the northern New South Wales country well. Mr Otway, Bennie Benson and I inspected one property at Walgett for agistment but turned it down when we found out there were pigs present, as pigs reduce the lambing percentage. It was a pity as it had good pasture.

A taxi driver took us out in a Ford car with loose steering and faulty tyres. It was a dirt road and after rain it had tracks six inches deep, so we jumped from one track to another. Mr Otway was in the front and Bennie and I behind. Bennie and I were saying how rough it was. Mr Otway heard us and said, 'What, are you chaps nervous?' Bennie reckoned the chief could have been a steeplechase rider if he was enjoying the bumps so much.

Mac Falkiner, Alec Cann and I were going from Sydney to Newcastle and the road ran parallel to the railway line. This day Mac was speeding along and about half-way up Mac said, 'This train is sticking to us today.' He said this because we heard the whistle every so often. Then a bit further along a police car came up beside us, which accounted for the noise and Mac was booked.

In Sydney, we had heard of a horse that was a good bet for the Sydney Cup so when Mac was being booked he told the two policemen it was a good tip. The race was the next day and it won, paying a good price; the charges were dropped. We went to the Quirindi races that day and we listened to the race near the bookies' stands. I thought it had lost and threw my ticket on the ground with other tickets. When I saw Alec he said the horse had won and he had collected, so I went back to where I had thrown my ticket and, miraculously, I picked it up.

I was good mates with a jackaroo called Victor Brennan, who was at Boonoke when I was there, but left after a few years and we heard he had started a stud. I had not heard of him again until a chance meeting about forty years later. Louise and I went to Farmers Store in Sydney to get a portable wireless for Louise, as she had just passed her Matric. We were dealing with a man who said, 'If you are in the country you will need a five-valve set. I was in the country once, up in the Hay area.' I said, 'What place up at Hay?' He replied 'Boonoke'. I then asked him what years he was there and he answered 1924–26. We still did not know each other until I said,

'Would you remember a jackaroo named Browser?' He then jumped the counter and grabbed me by the shoulders and said, 'Bowser, I thought you would be six foot four and thirteen stone by now!' I knew who he was when he called me Bowser because everyone else called me Browser. I don't know what other customers thought but we had a good yarn. He was a real character; even when he arrived at Boonoke, he was asked if he could ride and he replied, 'Yes, Mr Thompson, I can bump to a trot.'

Vic Brennan on Puckawidgee Bill

11 MOVE to WANGANELLA: the 1950s

Les Falkiner with Wanganella Estate sale rams

Gat's sister Louey and their mother
OPPOSITE: *Louey*

ABOUT 1950, THE FALKINERS SOLD the eastern portion of Moonbria, and Travis Falkiner, who was the manager there, then needed to move to Zara so we had to move, too. F.S.F. wanted me to go sheep classing or selling rams, which would mean living in Sydney or Brisbane. However, as we had a young family this did not appeal to me, so I accepted a position at Wanganella. We went there for three years and we were happy, until I went in the legs and broke down like an old horse. I eventually had to have a big operation on my spine and at the age of forty-six decided to settle in Deniliquin.

In the 1950s planes were being used for carting sheep about and when we were at Wanganella, I went out with Mr Austin to the ram house where Harold Munro was looking after the Sydney sale rams. Mr Austin said to Harold, 'We are sending the rams to Sydney by plane this year and we would like you to go there with them. What do you think of that?' Harold replied, 'No bloody good boss.' Mr Austin then asked 'Why?' Harold was in no doubt: 'Because if that plane crashes on the way to Sydney there will be headlines in the paper reading "Wanganella rams crash on the way to ram sales", but there will be not a bloody word about Harold Munro!'

Harold was like old Jack Walters, who was head stockman at Tabinga in Queensland. Mr Youngman had a friend who had his own plane and once when he flew to Tabinga, Mr Youngman said to Jack, 'Bill will give you a flight round Tabinga this morning. What do you think Jack?' Jack said, 'I have done what you have asked me for twenty-five years but I cannot go up in that plane, thank you.'

My parents had moved to a house overlooking Elizabeth Bay in Sydney in 1930 and they then moved to Deniliquin in 1952 with my sister, Louey, who had never married and always lived with them. They lived in a house in Wick Street, not far from where we live now. My mother died in 1956 and Louey met a man named Edward Le Souef and they decided to marry in 1958. Edward and Louey moved to Perth shortly after, taking my father with them. Edward came from Perth and he was a sheep-classer and came to the Riverina to purchase rams for Western Australian buyers.

12 CLAVERLEY

Ian Geddes, Dalgety agent,
Gat and Winkie

Fortunately, before we left Falkiners, Mr Otway and F.S.F. had sold us part of Zara: 3400 acres. It was funny how it came about as I asked Mr Otway one day if he would sell me a paddock. He said, 'Yes, Browser, you just write in to the office and I will put it to the board.' This I duly did and had the land valued by the bank manager and the manager of a wool firm in Deniliquin. I selected the southern end of Zara because it really gave Zara a straight southern boundary, too. After F.S.F. said they would sell it to me, they said I could pay for it when I could on those terms. They had to charge interest but charged the lowest.

So that's how we got our land, which we called 'Claverley'. We decided on that name after the area the Gatacres came from in Shropshire, England. When I saw Mr Otway later I said to him, 'That land that you sold me, I hear the irrigation is coming through and it will be worth a lot more money, did you know that?' He replied, 'Oh, that's all right Browser, you may as well have it instead of some other bugger. What I think is that instead of 3000 acres you should have got 5000 acres.' I was, however, very grateful to get 3400 acres in the end. When the irrigation came through, we got three miles of water commission channel, so we were very lucky.

Gat before his daughter Louise's wedding to Donald Officer

I worked those 3400 acres and later was able to expand on that. F.S.F. also had to fence off about 2200 acres of a 5000-acre paddock, plus another 1200-acre paddock to make up the 3400 acres. There was a three-mile fence there and the channel was on our side of the fence. They also gave me a brand-new Austin motor car as a bonus and it was a beautiful car. I purchased some Smeaton Vale-bred ewes at a Deniliquin sale on the advice of Dalgetys, and the Falkiners lent me Boonoke rams to join them.

We were lucky because the early 1950s were the height of the wool boom. In a few years, we won the Boonoke Cup. This was presented by F.S.F. to the flock owners of New South Wales with flocks of up to 2500 sheep. It was judged by the Department of Agriculture in different areas, and the winners of each area were then judged by the Chief Departmental Officer in Sydney, who came out for the occasion. We were lucky enough to win that trophy in 1952, which was good. We also won the Country Life trophy and the manager asked us what we would like as a trophy. Win said, a coffee table. They then wanted to know what kind so Win designed one and asked

Louise and Donald Officer about to cut their wedding cake

for a silver plaque to be put on it, saying what it was for. This was done and we still have the table.

In the early 1950s the irrigation water came through which allowed us to start growing rice, which was a very profitable crop. In those days, it was on an 8-year rotation. The area was then sown down to sub clover, which was very good sheep feed and put nitrogen back into the soil. With rice in a new area every year it allowed us to expand the area under irrigation. Initially we were only allowed to grow fifty acres; however, in the latter years the area planted was expanded. In the early days, we had a share farmer, Keith Owers, who came from Leeton where they have been growing rice since 1924. He also share-farmed for a couple of neighbours. This continued on until 1965 when Philip started to grow our rice and share farmed the rice on Zara and Wanganella

Our elder son, Philip, had decided when he left school that he wanted to go on the land so he used to come out and help me look around Claverley. Then in 1965 he took over the growing of rice. He also began share farming for F.S.F. and did contract work for them. When Sue and Philip decided to get married in 1967; we built a house on Claverley for them and they moved out there after their wedding in 1967. Philip then bought a block within half a mile of Claverley and decided to grow rice there. We have since increased our land holding by purchasing neighbouring land to the south of Claverley, including two good woolsheds and the Box Creek running through it. Philip now looks after all our land for us.

Our younger son, Michael, worked around Deniliquin for a while after leaving school but then went to the Northern Territory working on a cattle property, 'Brunette Downs', owned by King Ranch Pty Ltd. Then to the Kimberley to another cattle property, 'Mulla Bulla' owned by AMP. He finished up doing contract work and bull-catching up there for a period. When he got that out of his blood he came back to Deniliquin and is now manager of 'Woorooma Station' at Moulamein.

Our daughter, Louise, married Donald Officer and went to live in Melbourne. Donald's father, Don, was a great friend of mine in my early days at Boonoke.

Win and I live in Deniliquin still and are fortunate enough to see our family and grandchildren, too, who give us a lot of pleasure and interest.

I was on the Pasture Protection Board from 1961 – 75, including some years as chairman. I also did some sheep classing and ram selection for friends.

I often mentioned to the family about making dampers up in Queensland and how easy it was. The way I was taught, firstly you got three double handfuls of flour, which you put in a tin dish. Two dessertspoons of cream of tartar and one tablespoon of soda, which you stir up well. Then sprinkle that over the flour and add a little water to mix it, but don't make a wet mix. That is then put into a camp oven. This was a bit of a joke with the family. On my eightieth birthday, I went down to my son Michael's place with the rest of the family. He had the camp oven and had made a damper especially for me.

Claverley is still in the Gatacre family and is owned and run by Philip's son Jon and his wife Victoria and their children James and Charlotte.

Gat's eightieth birthday, L to R: Winkie, Michael, Gat, Philip and Louise

13 THE FALKINER EMPIRE: 1878–2000

Boonoke North stud rams, 1926

THE FALKINERS, LIKE A LOT OF BIG FAMILY BUSINESSES, had to sell in 1971. If Mr Otway Falkiner had kept fit, he may have been able to crack the whip and keep the families together. People get old, however, so this never happened.

After the Second World War the Falkiner family became fragmented: Otway lost a son John in the war and also a grandson. Ralph, his brother, died in 1946, and Una his beloved second wife in 1948. Leigh, the youngest, died in 1952, Norman having died in 1929 in London while receiving medical treatment. This left Otway by the mid-1950s the only one of the five brothers remaining and with no obvious successor.

Franc Falkiner started the Falkiner empire when he was able to purchase Boonoke and half the Peppin sheep stud from the Peppins in 1878. The other half was bought by Austin & Millear who also purchased Wanganella Station. They subsequently dissolved their partnership: the Austins retained Wanganella on the west and the Millears took the eastern half, which they renamed Wanganella Estate. The Falkiner family acquired Wanganella Estate in 1910 and Wanganella in 1958, therefore reuniting the Peppin flocks under the one ownership again.

In the early 1900s Franc Falkiner owned Boonoke, purchased in 1878 and followed by Moonbria in 1884, Tuppal in 1891, Moira in 1899, Perricoota in 1899, Warriston in 1909, Widgiewa in 1909 and Wanganella Estate in 1910. In 1917, 76,000 acres of neighbouring Puckawidgee was purchased and incorporated into Boonoke, thus forming the property as we know it today. In 1932, 50,000 acres of Widgiewa was sold to Herbert Field, leaving the homestead and 18,000 acres, which was named Boonoke North and was where Otway lived to run the empire after his father died in 1909.

Franc had formed F. S. Falkiner & Sons in 1899 and it has retained its name throughout the years, even when it was bought by Cleckheaton in 1971, then sold to Rupert Murdoch's News Ltd in 1978. Murdoch held on to it for twenty-two years during which time they added Barratta to the west of Zara. They spent millions of dollars improving the irrigation layouts, which have basically drought-proofed all the properties. They also spent a lot of money refurbishing the Boonoke Homestead. News Ltd finally sold all properties under the F.S.F. banner to Bell Securities in 2000 and they are still the current owners.

The Riverina properties under F.S.F. ownership

Boonoke	43,204 hectares (0.404686 hectares in 1 acre)
Billabong	648 hectares
Warriston	2,877 hectares
Peppinella	96,32 hectares
Wanganella	11,389 hectares
Zara	20,700 hectares
Barratta	31,000 hectares
Boree	100 hectares

Five Boonoke sale rams, 1925

14 MEMORIES of the FALKINERS

Mac Falkiner, Gat and Les Falkiner
at Jum Falkiner's wedding

Mr Otway

LOOKING BACK, I find I have some fond and amusing memories of my associations at Boonoke, especially with all the Falkiner family. Mr Otway Falkiner, in particular, I remember. He was an outstanding character in every way: firm, a good boss (everyone respected him as a boss), a generous man and good company. I knew of no family more generous than the Falkiners. All in the district who knew Mr Otway, knew him as a character and liked him. In Sydney, after a day at the rams' sales, he said, 'I want you to come with me to the Old Geelong Grammarian Dinner.' I said, 'I can't do that, Mr Falkiner, I didn't even go to Geelong.' He wouldn't be put off. 'Of course, you can; you can come with me.' We had a good and interesting dinner. Then he said, 'We will now go to the wedding (my memory of the couple hasn't stayed with me).' I again said, 'I cannot go there Mr Falkiner,' and Mr Falkiner again said that I must come with him. So, we went and enjoyed the reception which was at Elizabeth Bay House. Mrs Falkiner was already there.

Otway Falkiner and friends at Deniliquin Picnic Races. Otway Falkiner behind the wheel, Stephanie Barton and Beatrice Young, later Mrs Leigh Falkiner

Sometime in the evening the chief said, 'Mrs Falkiner wants to go back to the hotel so we will go.' We went to the entrance with Mrs Falkiner, where the chief's private chauffeur was waiting. Mrs Falkiner got in the car and the driver moved away and left the chief and me standing there. We decided then to return to the party, but Mr Falkiner said that this would take some explaining. Another time at a big party in Sydney champagne was the drink. Mr Otway was carrying a tray of about fifteen glasses when someone bumped him, knocking off all but three. He then tipped them off, too, saying, 'You may as well go with the rest.'

When Mr Otway won the Sydney Cup with his horse David in 1923 he went back to the Australia Hotel and the receptionist said to him, 'You had a good day, I wouldn't mind what you just have in your pockets.' With that he emptied his pockets and said, 'You can have it.' Otway collected about 120,000 pounds from the bookies equal to about $4 million in today's money, one of the biggest plunges in racing history. It was equal to three thousand suburban houses in 1923. (He later told us the receptionist built herself a house and said she didn't even ask him to see it.)

A story is told about Mr Otway at a local meeting. He had a horse running that was a certainty and he heard that his jockey was going to pull it. So, he told the jockey if he didn't win the race he would wring his bloody neck and that he had better keep going back to Echuca.

When Otway Falkiner was young he would ride around the back of Boonoke and when he returned his father would ask him how the cotton bush was in certain paddocks. If he said it was looking trimmed or 'it's trimmed back a bit', his father would say, 'Ottie, you'd better lighten that paddock off, there's too many sheep there.' So, he judged the carrying capacity and the way the sheep were doing by the way the bush was, salt bush and the cotton bush, but mainly the cotton bush.

One day Mr Otway had been working in the office all day and as usual he would get a bit bored with office work and go for a walk later in the afternoon around the homestead at Boonoke North. It was like a little township, with a school, carpenter's shop, barracks,

THE FALKINER FAMILY IN 1939

Front Row (Left to Right): Mr. FRED KNIGHT, LL.B. (Grandson of the late F. S. Falkiner), Wing Commander, R.A.A.F.; The Late RALFH S. FALKINER (Son of the late F. S. Falkiner), late Director; Mr. OTWAY R. FALKINER (Son of the late F. S. Falkiner), Chairman of Directors, General Manager; Mr. LEIGH S. FALKINER (Son of the late F. S. Falkiner), Director and Secretary; Mr. JOHN CARSE (Grandson of the late F. S. Falkiner), Director, Manager of Moonbria Station.

Back Row (Left to Right): Mr. BILL FALKINER (Son of the late Ralph S. Falkiner), served with A.I.F.; Hon. O. McL. FALKINER, M.L.C., (Son of O. R. Falkiner), Director, Manager of Boonoke Station; Mr. TRAVERS FALKINER (Son of late Ralph S. Falkiner), Flying Officer, R.A.A.F.; Mr. CHARLES L. FALKINER (Son of O. R. Falkiner), Flight-Lieut., R.A.A.F., Director; Mr. FRAZER FALKINER (Son of late Ralph S. Falkiner), Flying Officer, R.A.A.F.; The Late JOHN FALKINER, B.A. (Son of O. R. Falkiner), Flight-Lieut., R.A.A.F. [Killed in action]; The Late LEIGH FALKINER (Son of Leigh S. Falkiner), served with R.A.F. [Killed in action].

In Front: FRANC FALKINER (Son of C. L. Falkiner; Grandson of O. R. Falkiner).

Falkiner family, 1939

blacksmith's shop and hut. On his way around he looked in at the blacksmith's shop and saw Jonny Macree sitting on the anvil and Billy Wilson, the blacksmith, over at the forge working there. He thought Jonny didn't look very happy on the anvil so he said, 'What's going on here, Billy?' 'Oh,' said Billy, 'I was just pulling a tooth out for Jonny and I broke it. Now I'm drawing this pair of pliers out so I can get out the root.'

Mr Otway said, 'Forget about it Billy, you come with me, Jonny.' He took him back to the house, gave him a whisky and then took him to the dentist in Narrandera.

'Jum'

Fraser 'Jum' Falkiner also became a very good friend of ours after coming to Boonoke in 1938 on finishing school in England. Little did he know that he would be back in England in early 1941 flying Spitfires over Germany. Jum was the first pilot to fly a Spitfire in an operational role.

Jum did his initial training in Australia before being posted to Canada in December 1940 where he got his wings and was then sent to England where he had his twenty-first birthday just a few weeks before he was shot down over northern France. His plane was on fire and his canopy release was jammed so he could not get out. However, he was able to force it open by planting his feet on the instrument panel; he free-fell about twenty thousand feet. When he came to and had the presence of mind to pull his rip-cord, he landed in a field with a badly burnt face; he spent seven months in hospitals before being imprisoned for three and half years in German POW camps.

After Douglas Bader lost his tin legs while parachuting from his crashing plane over France, and ended up a prisoner of war in St Omer hospital, Jum was one of the escorting Spitfires to a Blenheim bomber that dropped a new set of legs on the St Omer airfield for Bader. It was one of the hospitals that Jum had also spent time in.

Mac and Betty Falkiner with Gat

F.S. Falkiner

The Falkiners also owned a property called Tuppal, where they used to shear over 150,000 sheep annually and in the drought year of 1902 they shore 207,515 for a total of 3444 bales. Mr Bert Falkiner used to manage Tuppal and when it was sold he saw Mr Otway who was at Boonoke North and he said, 'Ottie, I don't know what to do now as there's no jobs with the firm, I must buy a place.' (And that's just what he did, he bought Haddon Rig, which has also become successful.) He said to Ottie though, 'There's one thing I'd like you to do. I would like you to look after my teamster for me, old Jonny Macree; he's been with me a long time, he's a good man and I would like you to give him bedding for the rest of his life and a glass of scotch whisky each night.' This happened and in those days, they used to get their whisky from Melbourne in a forty-four-gallon container, so old Jonny got the dinky-di stuff.

Betty and Mac Falkiner, John Wilson, Gat, Bob and Pam Sefton with their daughter Prue and Winkie

George

Bert Falkiner's son, George Falkiner, was a keen pilot and had two aircraft, a Waco and a Stinson. He also had a sixty-foot luxury cruiser on Sydney Harbour and was in business with the famous war-time flying ace Clive (Killer) Caldwell. Caldwell also happened to be my brother Galfrey's best friend; in fact, he delivered the eulogy at Galfrey's funeral.

Clive Caldwell was the leading Australian air ace of the Second World War. He is officially credited with shooting down twenty-eight enemy aircraft in over three hundred operational sorties.

Les

Mr Les Falkiner was returning from the Echuca Picnic Races one year in the drought when there wasn't much water around. There were a lot of swaggies about at the time and one of these swaggies was on the side of the road. There used to be a fence along the road and there was a gate on the side of the road to get into Moira. The swaggie was at the gate and he opened it for Mr Falkiner, who drove his old Rolls Royce through. He stopped and thanked the swaggie who asked if he had any water. Mr Falkiner replied, 'I haven't got any water but I can give you a bit of liquid though.' He leant into the back seat, where he had a case of champagne and grabbed a couple of bottles for the swaggie. I think the swaggie thought it was Christmas.

Swaggies

In the depression years most travelling swaggies were looking for a job, but there were some who were just tourists. They would call at the station and see the station cook and get a handout, enough food to keep them going to the next stop. For example, they would call at Boonoke, then the next stop would be Wanganella. The cook would usually give them some flour, bread (if cooked at the time), cooked meat, tea and sugar. The swaggies knew all the cooks and they had them all named, such as one who was called 'Raw material'. This

Friends at the Deniliquin races,
Pat Gorman, Jack Webb, Cyril Gove,
Theo Macaw, Neville Armytage
and Bob Bradshaw

was because when they asked for a handout, he would always say, 'I'm sorry, I have got nothing cooked for you today, mate, but I can give you the raw material.'

They used to follow the rivers and watercourses. The North-West paddock on Boonoke was called 'five to four'. It had a government telephone line running across it, heading towards Hay. It was told that the reason it was called 'five to four' was that somebody died when he was walking along that telephone line. When they found him dead his watch had stopped at five to four.

Les Falkiner had the captain of the HMAS Renown up for the Picnic Races once and he took him out kangaroo shooting in his Rolls. Les drove fast and the Rolls had to go through swamps and rough ground. He said they were good shooters because they stood up shooting and had their sea legs because they did not mind the bumps.

Cyril Gove

It was later at the races Cyril Gove was walking to the bar with the captain for a drink. Cyril Gove was a good local amateur rider, as was Swanee Edgar, the manager of Wanganella Estate. At the bar, Cyril asked all to have a drink as the Renown was shouting. One could see the captain noting with his eyes how many, so when paying-up time came the person next to him said, 'Drink is free today.'

Cyril was known for riding winners, from Riverina to the Western District of Victoria. He played League Football with Essendon, representing his team in the state side, and was also an amateur boxer. It is reported that in one day he rode in a horse race at Moonee Valley, jumped into a taxi and hastened to Essendon to play in the football match of the day, and then won an amateur boxing match in the evening. A good story is told of his being approached by a representative of Who's Who with a view to having his name included in its pages. Asked, 'To what creed do you belong?' thinking the question was 'What creek are you on?' he answered, 'Two – the Colligen and the Tumudgerie...'

There is another story I remember, involving Cyril Gove. He was in Deniliquin for a sheep sale, and after the sale decided to catch up with his friend Bob Alley for a few drinks. After some time Bob asked Cyril to come back to his home for dinner; Bob's wife Londa greeted both of them. Bob had not told Londa that Cyril was coming for dinner so when Bob and Cyril sat down Londa gave Cyril his meal and then sat down with hers. Bob asked Londa where his meal was and she replied, 'Cyril's eating yours.'

Cyril Gove and Swanee Edgar, after one Picnic Race meeting, collected all the jerries (china chamber pots), which were still put under beds or in the cupboards under the wash basins. As a joke, they got all the pots out of the bedrooms at the Royal Hotel and hung them on a rope across the street to the Exchange Hotel. The police did not approve and said they had to be removed immediately, so Swanee took his knife and cut the rope. This passed without headlines.

Dick Nicholson

Dick Nicholson, who was my offsider in the 'thirties, told me a story about himself and Shore Moffat riding Mac Falkiner's horses. They used to do track work with them for the Southern Riverina Picnics and would be at the track by daylight to do their riding work. Then afterwards they would drink rum and milk at the Globe and be back at Boonoke for orders at 7 a.m. Another time Les Falkiner took Commodore Holbrook to the Hay races and returning to Zara he decided to come the short way and got bogged at the gate to the black swamp. He and the Commodore walked across the salt bush to Cooks Ridge outstation at about 3.30 a.m. Holbrook said it was very good navigation at night. He knocked on the cottage door and Harold Munro said, 'Who's there?' Les said, 'Mr Falkiner, Harold.' Harold, not believing it, said, 'Mr Falkiner, I'll be buggered.' So, Les had to say, 'No, it is Mr Les, Harold.' Harold then came out with a hurricane lamp and put it up to Les's face and said, 'Oh, it is you, boss.'

A story was told of an old manager at Barratta who walked into the office after riding all day in the rain and looked at his barometer. It said 'fine' so he pulled it off the wall and as he threw it out the door addressed it, saying, 'Now see for yourself!'

*Winkie, Gat in the background
and Otway Falkiner at
Jum Falkiner's wedding*

15 YARNS FROM MEMORY

Boonoke lamb-marking team, 1925:
Reg, Roy and Birdie, Billie McKees,
Robbie, Jack Edwards and Vic Brennan

THIS STORY I THINK I HEARD when I was at school. A chap came in from Charleville, where there was a hotel which was run by the Marconis. A lot of the judges from Brisbane had been there duck shooting. They had two or three days' duck shooting and then they returned. One of them left his gun behind, so the publican rang up this judge and said, 'You left your gun behind.'

The judge asked, 'My what behind?'

'Your gun, you left it in your room.' When the judge asked: 'Well, spell it' the reply was: 'G for Jesus, U for onion and N for pneumonia, GUN.'

Jackaroos' fun

Jackaroos from Moonbria, Boonoke, Wanganella Estate and Zara would combine monthly for the F.S.F. jackaroos dinner. One such night there was a dinner at the old Royal Hotel. There were apparently thirteen or fourteen around a long table. I was overseer at the time at Boonoke and although I did hear from Mr Thompson the next

Gat having lunch in a paddock, Claverley, 1960

morning about it I was later phoned by the police. The Sergeant of police said that he had a jackaroo in gaol and he would be bailed out. I was told the evening started off quietly until the jackaroo, pouring out the beer from a jug, decided to liven things up. As he walked around the table he poured some beer down one poor jack's neck; then it really started things going.

In those days, the police had the land where the water commission is today and it went back to Cressy Street, with a high cement wall round the Cressy Street side and part of Edward Street. Another jackaroo was put in for the night for some minor offence and he escaped out of his cell and climbed over the wall and jumped into Cressy Street just as two police came around the corner from Edward Street so was put back again. That same boy who was a New Zealander, ended up in the Changi camp but the poor fellow didn't escape from there.

Another time they had a party at some hall in Deniliquin and at that time smoking was not allowed. One boy was smoking there and the policeman asked him to put his cigarette out and he refused to put it out immediately. The boy was then reprimanded for smoking and said, 'Oh yes, I know you, you're the one they call guts and buttons!' With this, the policeman took him by the arm straight to the police station. You could not blame the police.

Doctors' stories

There were some sad stories of men who couldn't cope with life in the 'thirties. I recall Doctor Gorman was called out to a property once, where an owner had chosen the woolshed as the place to shoot himself. Dr G. found a chap lying on the floor near a bale of wool with a shotgun at his side and a hole in his head. He had fastened a note to the wool bale and then shot himself in the mouth, blowing his brains out. His brains had hit the note on the bale and therefore the note was unreadable.

Another chap had ridden to the doctor's surgery and complained of head pains and worries. The doctor spent a long while talking to the fellow and he thought he had helped the man. However, on his way home, the man had another fit of depression. He rode his horse under a tree limb on the Warbreccan Reserve, attached a rope to it and made a slip knot round his neck; he then faced his horse towards home, gave it a kick with his heel and that was the end of him.

There were some amusing times, too, as the doctors dealt with all sorts of people and problems. The same doctor was having a few drinks with four or five of his male friends, including one who slurred his words badly after a few beers. A chap, who had a cleft palate, came in to see if the doctor would cure his sick cockatoo. So, the doctor said, 'Just wait a minute, I have a chap here who knows all about cockatoos,' and they sent out the friend who was slurring his words. The doctor and his friends were inside listening to the two trying to make one another understand.

A sixteen-year-old boy went into Doctor Middleton, another doctor at the time, with a stomach pain. The doctor said, 'You have appendicitis and you need an operation.' The boy asked if he would do it, but what would he charge. The doctor said, 'I charge two, four, ten or fifteen guineas, according to what you can afford.'

'The boy said, 'Well, how about two guineas?' The doctor agreed and took his appendix out for what he could afford.

Watery stories

When F.S.F. owned Moira, its lakes were renowned for snakes; snake-catchers would go there especially for them. Sitting around the camp fire at night when out lamb marking, one would start up a talk about getting snakes, etc. When the tales were good, one would throw a dead snake among the men to see who was afraid.

One experience I had myself was when I was about to fill my quart pot in Browns Creek. I noticed colours under the water and thought of a tiger snake, sure enough it was lying under about six inches of water. I then wondered how long a snake could stay under water. Another day I was about to boil my quart pot on a dam and saw a dopey stumpy-tail lizard. I pushed it under water at the edge of the dam, boiled my quart and had my lunch. I then raked out the lizard and it was no worse from being an hour under water.

A chap at the outstation was setting off one morning and was cleaning out the troughs at a mill near the horse paddock, when a big brown snake wriggled into a two-inch pipe lying on the ground. He blocked up both ends and put it into a trough. Returning that evening, he shook out the snake thinking it would be dead but it was livelier than ever.

A good milkman/butcher was as hard to get as a good cook. We had a milkman who was putting water in the milk so I said to him, 'The milk is getting watery, Jack.' He replied, 'Well Mr Gatacre, the cows did swim the creek this morning.' So, I said, 'Well, from now on you had better keep the cows away from the creek!'

Once I remember, we had two women helping in the house and sometimes they would walk to the weir on the creek and dangle their legs over the apron, in the water. When yabbies bit their toes, they would push a piece of netting behind and under them and lift them out. There were plenty of yabbies in the creek and they caught quite a lot of them. They made good eating.

Bush dentistry

They were tough in those days. I can still remember when we were camped out once while lamb marking at Boonoke and we had a chap there who was a station hand but also a shearer, horse trainer and other things. He complained about a tooth hurting and that night asked who had a pair of pliers and would they pull the tooth out for him. Someone had a pair so he sat down on the bed and said, 'Go on, pull it, it's been aching all day.' The chap pulled the tooth out with his pliers and the station hand just went and rinsed his mouth out with a little water. He was back lamb marking with us the next day.

Another story about a tough man was that of Scotty McMillan, the blacksmith. He was in our yard team sometimes when jetting and he looked after the engine and mixer and mixed up the sheep dip. He only had one eye but managed well with that one. He reckoned he lost his eye when sighting the line of a fence and a big grasshopper flew into his eye. Anyway, one day he was cranking up the engine after lunch when the crank handle flew off, hitting him in the mouth. Scotty spat out three teeth and said, 'Strike me bloody lucky,' and then carried on. He used to make his own horehound beer in his blacksmith's shop by boiling up wild horehound. It was said he would put a bit of spirits of salts in it too, to give it a kick.

Wisdom on the run

Mr Thompson and Mr Falkiner were driving around one day and they called at the Boonoke shed, which was being built at the time. The chief saw something wrong and tore strips off the contractor. When driving away Mr Thompson said, 'You were wrong rousing on that chap, boss, it was not his fault.' Mr Falkiner replied, 'Well, if he didn't deserve it today he would another day.'

One day at Boonoke, out with Mr Thompson, we checked on how the rabbit camp was getting on at the back of Puckawidgee. The camp had a character, George Durden, in charge of six or eight blokes, who were busy digging out rabbit burrows. Mr Thompson said, 'Don't you even give them smoko, George?' At smoko you

ABOVE: *Blacksmiths shop*
OPPOSITE: *Sheep ready for drafting, Boonoke, 1924*

will wait five minutes and you will think it is the Deniliquin train, as there will be so much smoke (from everyone smoking). Later, George and Jimmy Williams were fencing. It was hard black soil and Mr Thompson and I drove up. Mr Thompson said, 'It's pretty hard, George, but one thing, if you and Jimmy have a row, you'll have plenty of ammunition.' George replied, 'Ammunition be blowed, a man'd be too weak to throw it. Last night walking back to our camp, I thought Jimmy was pulling me back but it was only my watch had fallen out of my pouch and was dragging on my belt.'

DAVID GOVE'S MEMORY

Gat with Bill Lamb, Principal of Willurah,
at their field day

159

Mr and Mrs Gatacre and
their daughter Louey

MY MEMORIES OF MR AND MRS GATACRE senior and their daughter Louey, are only as vivid as an inattentive teenage youth of the 50s would allow to be retentive.

My brother and I were taken to the Gatacres' house at 15 Wick St on several occasions, mostly as I recall for afternoon tea. I remember being greeted by two delightful people, always impeccably groomed, Mr Gatacre always wore a tie, whatever the weather conditions were, and our mother insisted we both had ties on when we went to the Gatacres, although we were most likely adorned in ties when we came to Deniliquin.

There was always a beautifully laid-out afternoon tea, with fine English china cups, saucers and plates, with a range of plain but tasty fare, such as scones or drop scones, set out in the back room of the house.

From memory you walked down a central passage to the living area adjacent to the back garden. In this room my very vivid memory is of the most outstanding collection of spoons, contained in a beautifully designed cabinet with an opening glass door. These spoons I suggest were all Sterling Silver, and from a great array of countries, signified by their coloured emblems.

I remember Mr Gatacre saying they were staunch members of the Church of England, but Browser, whom he always called Melmoth, had married a Catholic, but while it did sink into my immobile brain, it did not reverberate as something that needed retention as a matter of importance.

Now to their daughter, Louey; she was the epitome of elegance, with an electrifying personality, who loved her Dalmatian dogs, and they were a permanent part of her travel in her Austin A 40 car. Sometimes one wondered whether the dogs were driving or Louey was at the wheel.

My late Uncle Robert Vivian Gove, Chairman of the Melbourne Racing Club, said Louey was one of the most enchanting women he had ever met, and that heralds what so many people thought.

Along with her dogs and her personality, she made artificial flowers, and I can remember her taking these highly crafted products into Miller's Department Store in Cressy St, and taking them to Miss Singue who ran the Haberdashery section of the store.

During these interludes, the dogs remained in the car, and from memory there was an older dog, not sure if it was a male or female, and a younger one that was more playful.

While I was not living in Deniliquin at the time, I can remember my parents shock at Louey becoming engaged to Eddie Le Souef.

Eddie was a relation of Tom Eastman's and he described Eddie as 'Odd'.

These are a collective of some very vague memories of a gracious and delightful family, whose tradition and elegance were a hallmark of what my mother would call 'Good breeding'.

Reginald Henry Winchcombe and his wife, Christian Esson Gatacre, Gat's father and mother

APPENDIXES

Rice harvesting on Claverley, 1980

Gat and Winkie, 1990

Gatacre of Gatacre: Shropshire landowners for 700 years

Leslie G. Pine

Gatacre of Gatacre, co. Salop; a family seated at that place, since the time of Henry III, which lands were held of the Crown by military service, and acquired originally by grant from Edward the Confessor.

Thus wrote Sir Bernard Burke, and if we were to listen to him unreservedly, we should have to believe that the county families now existing were all of Norman origin, or – contradictorily – that the Conquest of 1066 had made little difference in the composition of the landed class. Perhaps the Gatacres did originate in pre-Conquest England.

Shropshire was not the easiest of counties for the invading Normans to settle; Welsh neighbours with a propensity for roving into England on armed forays may have rendered the continued occupation of an estate by a strong Englishman, tolerable to the Normans.

By the time that the first Gatacre in written records appears, he is indistinguishable from the rest of the knightly class.

Stephen de Gatacre in the reign of Henry III (1216–1272) possessed the manors of Gatacre and Sutton, with lands in Claverley, which he held of the King. This Stephen Gatacre was hardly likely to have been a new man. His ancestors could have been at Gatacre for 200 years before in the reign of Edward the Confessor (1042–1066).

Of other claims to fame, the Gatacres can be certain. Like many of the great country families they owe their surname to the lands which they have held – as far as written records trace them – for 700 years. The senior representative lives in Canada, but Gatacre Hall, Bridgnorth, remains the seat, under Elizabeth II, as under Henry III.

Like that of many other of our oldest recorded families the Gatacre coat of arms is simple. Quarterly gules and ermine, in the 2nd and 3rd quarters three piles gules, over all on a fesse azure five bezants. The crest is a raven, and the motto appropriately Hic eram in dierum seculis (I was here for generations).

Like many others, too, of our proudest families, the Gatacres have resisted the temptation to possess any title beyond a knighthood, reminiscent of the great Breton family's boast: 'Roi?, noble? Rohan, je suis.'

One of the most difficult things for the continental nobles or the modern British parvenu to understand, is the indifference of our oldest houses to the lure of the peerage. It is enough to be Gatacre of Gatacre. Incidentally there is probably no Salopian family of any eminence with which the Gatacres at some time have not intermarried, so that their pedigree is the perfect research ground for all those whose families are of ancient Salopian descent.

Not that the family was without dangers to face. Holding land of the Crown and loyal to the King, meant that the King's tenant-in-chief had to serve the King on the numerous warlike expeditions of the Edwards and the Henries.

What if there were doubt of the King's identity? In the reign of the unlucky Henry VI, Humphrey Gatacre was Esquire of the Body to that monarch. He lived until 1509, long after Henry's death. Perhaps Humphrey, being a younger son with little to lose, escaped, despite his attachment to the Lancastrian King. Perhaps 'treason doth never prosper, for if it do, none dare call it treason.'

Humphrey's elder brother, John Gatacre of Gatacre, was bailiff and M.P. for Bridgnorth in 1471 in the thick of the Wars of the Roses. However, the family continued in undisturbed possession of their lands.

At the time of the Reformation, a divide opened in the family. Francis Gatacre was in the Indian Mutiny, the China War, in Afghanistan and Burma.

His brother, Major General Sir William Gatacre, was in the Hazara Expedition, imprisoned under the first Elizabeth as a Recusant in 1575–76, and again fined as a Recusant in 1598. His youngest son, John, was sent to the Catholic College at Douai in Belgium from which came the stream of seminary priests who endeavoured to recall England to the Catholic Faith.

John, however, declined to become a priest, and it may be that he was not reluctant to be captured by Sir Edmund Uvedall at Flushing while trying to return to England. John Gatacre afterwards conformed to the Anglican Church, and was sent by the Privy Council to a university in 1595 (at his father's expense, the latter remaining a hardened Recusant).

Meanwhile Francis Gatacre's younger brother, the Rev. Thomas Gatacre, had become a learned Protestant divine and founder of the Mildenhall line.

Life was more settled for the Gatacres in the next generation. Francis's son, William, was appointed Cockmaster to King James I in 1607 with an annual fee of 100 marks, the equivalent £66.13.4.

Then follows a succession of Gatacres of Gatacre in peaceful possession of their ancestral lands, Deputy Lieutenants, Justices of the Peace and High Sheriffs of the county and Colonels in the local militia. In the past 200 years generations of soldiers have carried on the traditions of the family. Major General Sir John Gatacre served in Burma, Chitral, the Nile Expedition, and the South African War. Two of his sons were in the First World War, one was taken prisoner, the other was killed in action in 1914.

Captain Edward Gatacre of Gatacre, the head of the family, died of wounds received in action in 1916.

Turning to the junior line of the Gatacres, we begin with a succession of clergymen, all three of whom have made good their entry in the Dictionary of National Biography. The Rev. Thomas Gatacre mentioned earlier was an eminent divine educated at Oxford and at Magdalene College, Cambridge.

About 1553 he was a student at the Middle Temple; he became domestic chaplain to the Earl or Leicester and in 1572 Rector of St Edmund the King, Lombard Street, London, where he remained until his death in 1593. He is mentioned in Fuller's Worthies of England.

His son, Thomas, began the spelling of the surname which is used in this branch, Gataker, thus illustrating the diversity of spelling combined with unity of pronunciation known to our ancestors. This Thomas Gatacre was a Puritan divine and (it seems unnecessary to add for a Puritan) a critic. He was a scholar of St John's College,

Cambridge, Fellow of Sidney Sussex College, Cambridge, in 1696, B.D.1603 M.A., lecturer of Lincoln's Inn, 1601, and Rector of Rotherhithe, Surrey, for 40 years.

He was an active member of the Westminster Assembly, but unlike many of his Puritan brethren he favoured a mixture of episcopacy and Presbyterianism, and signed an address against Charles 1st trial. He published several works.

He was succeeded in the representation of the family by his son, the Rev. Charles Gataker, who went to St Paul's School and Sidney Sussex College, Cambridge, where he became M.A. (1636). He was chaplain to the famous Viscount Falkland, killed at Newbury and became Rector of Hoggeston in Buckinghamshire

His inclinations were towards the Puritanic side of the Church since he published a book against the High Church Bishop Bull's treatise, Harmonia Apostolica; he also produced an Examination of the Case of the Quakers concerning Oaths. He died in 1680.

The Mildenhall estate in Suffolk came into this branch in the 18th Century.

Gat with his first cousin, Des Gatacre, from England

RMS Niagara, *the ship on which Gat's father returned to England after his father died*

Family tree

For further information in relation to Gatacre family tree refer to:
Burke's Landed Gentry, Edn 18, Vol. 3

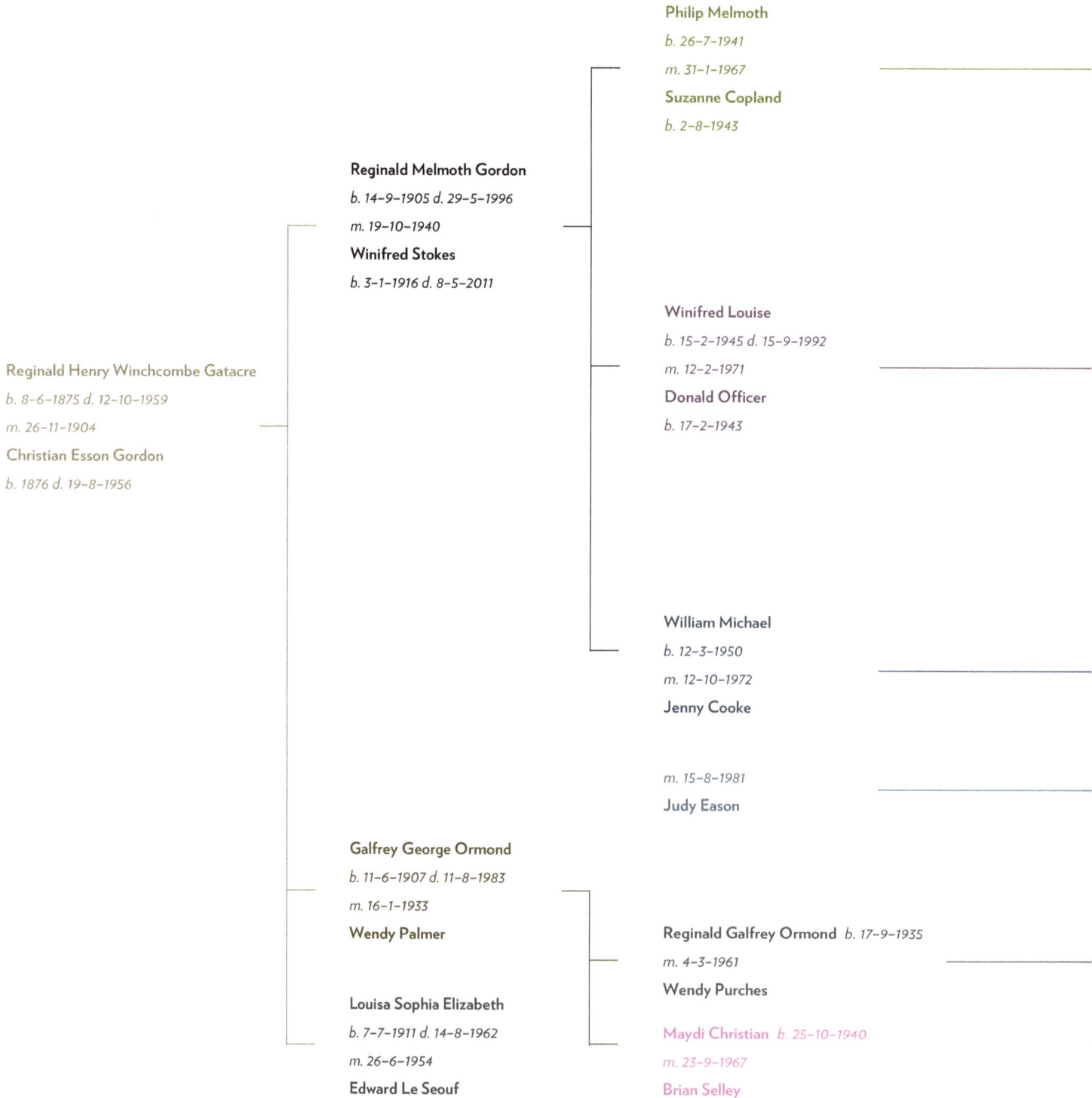

Philip Melmoth
b. 26–7–1941
m. 31–1–1967
Suzanne Copland
b. 2–8–1943

Reginald Melmoth Gordon
b. 14–9–1905 d. 29–5–1996
m. 19–10–1940
Winifred Stokes
b. 3–1–1916 d. 8–5–2011

Winifred Louise
b. 15–2–1945 d. 15–9–1992
m. 12–2–1971
Donald Officer
b. 17–2–1943

Reginald Henry Winchcombe Gatacre
b. 8–6–1875 d. 12–10–1959
m. 26–11–1904
Christian Esson Gordon
b. 1876 d. 19–8–1956

William Michael
b. 12–3–1950
m. 12–10–1972
Jenny Cooke

m. 15–8–1981
Judy Eason

Galfrey George Ormond
b. 11–6–1907 d. 11–8–1983
m. 16–1–1933
Wendy Palmer

Reginald Galfrey Ormond *b. 17–9–1935*
m. 4–3–1961
Wendy Purches

Louisa Sophia Elizabeth
b. 7–7–1911 d. 14–8–1962
m. 26–6–1954
Edward Le Seouf

Maydi Christian *b. 25–10–1940*
m. 23–9–1967
Brian Selley

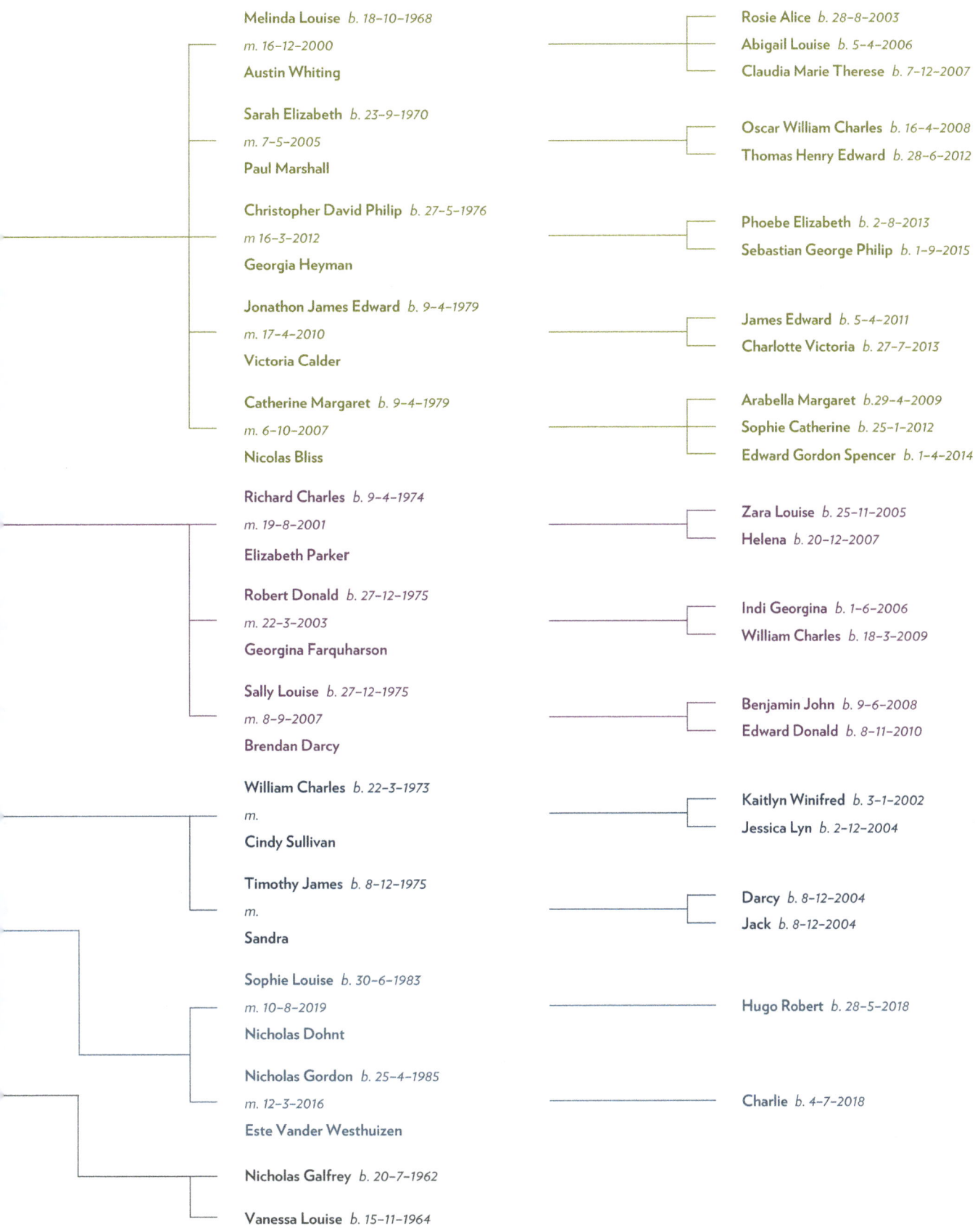

Melinda Louise b. 18-10-1968
m. 16-12-2000
Austin Whiting

Sarah Elizabeth b. 23-9-1970
m. 7-5-2005
Paul Marshall

Christopher David Philip b. 27-5-1976
m 16-3-2012
Georgia Heyman

Jonathon James Edward b. 9-4-1979
m. 17-4-2010
Victoria Calder

Catherine Margaret b. 9-4-1979
m. 6-10-2007
Nicolas Bliss

Richard Charles b. 9-4-1974
m. 19-8-2001
Elizabeth Parker

Robert Donald b. 27-12-1975
m. 22-3-2003
Georgina Farquharson

Sally Louise b. 27-12-1975
m. 8-9-2007
Brendan Darcy

William Charles b. 22-3-1973
m.
Cindy Sullivan

Timothy James b. 8-12-1975
m.
Sandra

Sophie Louise b. 30-6-1983
m. 10-8-2019
Nicholas Dohnt

Nicholas Gordon b. 25-4-1985
m. 12-3-2016
Este Vander Westhuizen

Nicholas Galfrey b. 20-7-1962

Vanessa Louise b. 15-11-1964

Rosie Alice b. 28-8-2003
Abigail Louise b. 5-4-2006
Claudia Marie Therese b. 7-12-2007

Oscar William Charles b. 16-4-2008
Thomas Henry Edward b. 28-6-2012

Phoebe Elizabeth b. 2-8-2013
Sebastian George Philip b. 1-9-2015

James Edward b. 5-4-2011
Charlotte Victoria b. 27-7-2013

Arabella Margaret b.29-4-2009
Sophie Catherine b. 25-1-2012
Edward Gordon Spencer b. 1-4-2014

Zara Louise b. 25-11-2005
Helena b. 20-12-2007

Indi Georgina b. 1-6-2006
William Charles b. 18-3-2009

Benjamin John b. 9-6-2008
Edward Donald b. 8-11-2010

Kaitlyn Winifred b. 3-1-2002
Jessica Lyn b. 2-12-2004

Darcy b. 8-12-2004
Jack b. 8-12-2004

Hugo Robert b. 28-5-2018

Charlie b. 4-7-2018

Map

Zara

Barratta

Claverley

km 0   2   4   6   8   10 km

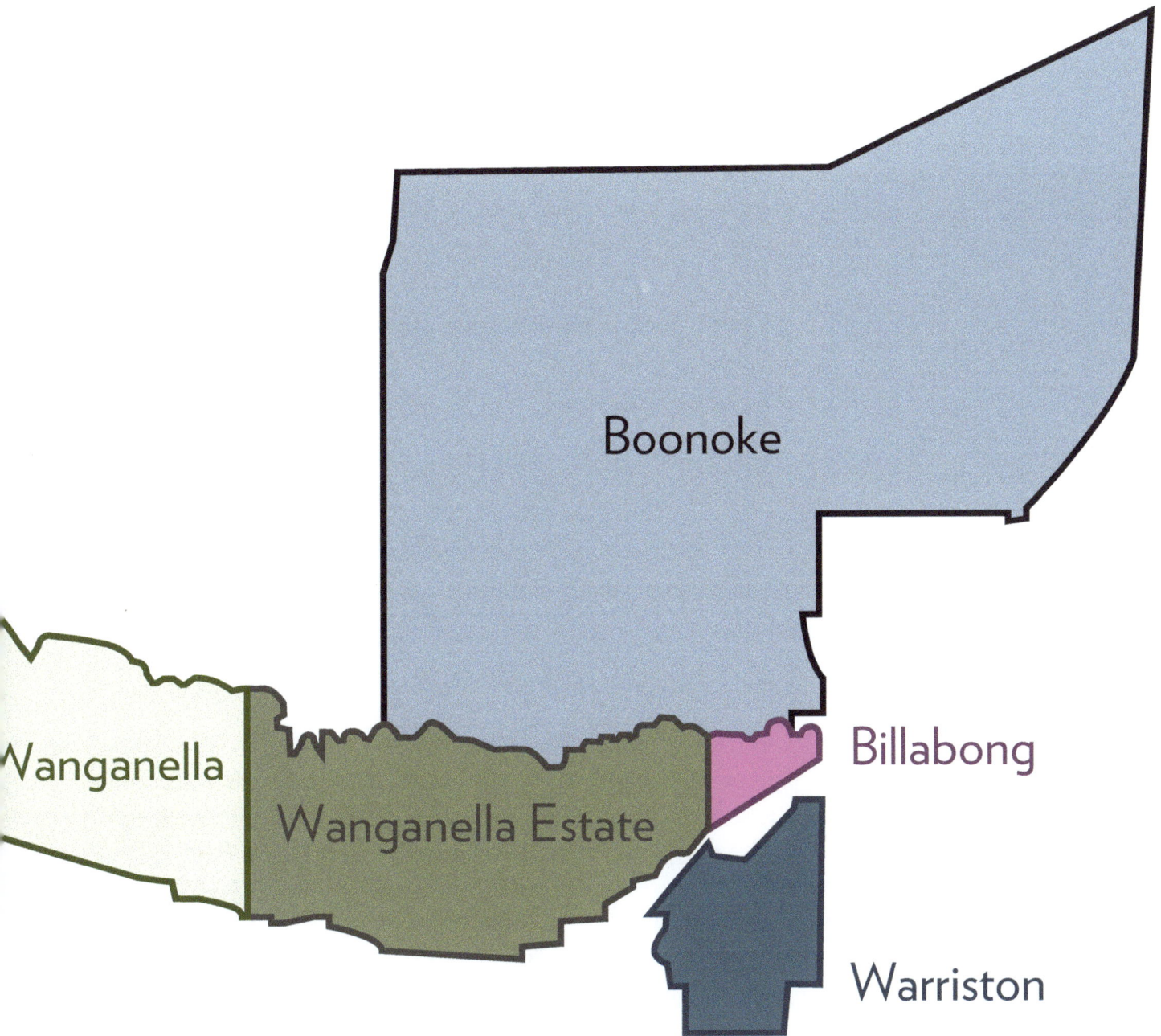

Boonoke

Wanganella

Wanganella Estate

Billabong

Warriston

Acknowledgements

I would firstly like to thank my daughter Melinda who in 1995 initially collated all of Gat's notes and recordings and transferred them onto a computer. It was a quite an effort trying to understand a lot of the terms Gat had used, some of which she had never heard. Then my daughter Sarah also spent many hours organising the chapters within the book. Thanks also go to my younger son Jonathon who has helped me complete the book, along with his wife Victoria who compiled approximately four hundred photos which have greatly complemented the book.

I would also like to thank David Gove for his knowledge and appreciation of Gat's parents.

Special thanks to Nan McNab who along with Bet Moore has spent a considerable amount of time and commitment in putting the book together. I was fortunate to have Angela Berry introduce them to me.

Thanks also go to Simon Horsburgh who has done an amazing job with the design.

Special mention to Mark and Naomi Ritchie for allowing Julian Atwell to scan all the photos at a high resolution and even repair some of them.

Finally I would like to thank my wife Sue who has done all the proofreading and also thank the rest of my family for their support and enthusiasm, which has helped make the book a reality.

About the Author

My father's ability to tell a story with humour and detail inspired this book. As his eyesight failed in later life we gave him a tape recorder with bandaids on the record button and the old stories flowed out of him as if he were reliving his past.

As Gat's eldest son my only responsibility was to collate all these stories and the history of his most interesting life and family.

I was born in 1941 when Gat was manager of Zara and when I turned eighteen I worked alongside him at Claverley. I learnt a lot from my father and realised from a young age how much knowledge he had and how admired he was.

This book has taken me quite a long time to complete with a lot of help from many different people. I hope you enjoy 'Gat' as much as I have enjoyed putting it together, especially the wonderful old photos, which are a great reminder of our past.

Philip Gatacre

Philip on Timmy, Zara, 1946

www.ingramcontent.com/pod-product-compliance
Lightning Source LLC
Chambersburg PA
CBHW040318100426
42811CB00012B/1475